THE SMART GUIDE TO

The Solar System

BY PHILIP SELDON

Cover design: Lorna Llewellyn
Back cover design: Joel Friedlander, Eric Gelb, Deon Seifert
Back cover copy: Eric Gelb, Deon Seifert
Illustrations: James Balkovek
Production: Zoë Lonergan
Indexer: Cory Emberson
V.P./Business Manager: Cathy Barker

ACKNOWLEDGMENTS

First, I would like to acknowledge my parents, long departed, who indulged me as a youngster with staying up late to observe the heavens and for giving me a telescope for my tenth birthday. When I was in my mid-teens, they permitted me to travel throughout the country to scientific conferences, which made it possible for me to be named a Director of the Smithsonian Astrophysical Observatory's Moonwatch Program at sixteen. They also didn't object to my moving to New York City on my own to oversee the artificial satellite observing station atop what was then known as the RCA Building, the city's second tallest building.

I would also like to acknowledge my long-departed mentors, Richard Luce of the Telescope Making Division of the Amateur Astronomer's Association of New York for assisting me in building my scientific instruments; to Armond Spitz, manufacturer of Spitz Planetarium Projectors, who provided guidance and encouragement; and to Richard Priest, a member of the Amateur Astronomer's Association who, in addition to sharing our interest in astronomy, introduced me to the wonderful world of wine, which became my career as founding editor and publisher of Vintage magazine, America's first wine magazine.

I want to thank and acknowledge Lesley Pratt for her assistance in creating this book. Her tireless efforts in researching, editing, and rewriting has been invaluable, and this book could not have been produced without her. I would also like to thank Kevin Marvel, Executive Officer of the American Astronomical Society, for his exceptional efforts in vetting the manuscript of this book while it was on deadline to ensure that it was up to date and did not contain any errors. A thank you to James Balkovek for his wonderful illustrations, to Zoe Lonergan for her production of the book, and to Cory Emberson for her proofreading and indexing.

Lastly, I would like to acknowledge and thank all the astronomers who have devoted their lives to the tireless research that has led to the discoveries that have made this book necessary. They are truly devoted to their field and work out of a love for their profession.

Philip Seldon

TABLE OF CONTENTS

PART ONE

Prologue

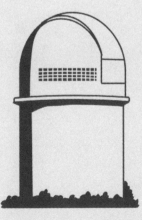

CHAPTER 1

Our Solar System

We are all very fond of our home planet, Earth. We feel secure standing on *terra firma,* the solid ground under our feet. But Earth, the third rock circling the Sun, is a special place in our solar system. In this chapter we'll talk about the different types of objects in our solar system (and there are many of them, with unique and sometimes stunning features) before discussing them in more detail in later chapters.

Thousands of years ago, people looked up at the night sky and saw thousands of bright, dazzling objects. They didn't have what we now call "light pollution"—the bright lights that mask most of the stars from view in our cities and suburbs and even the open areas that surround them. Gazing up at the blanket of stars, the people could see that the stars rose on one side of the sky and set on the other.

Some of the stars dazzled brighter than others. And while most of the stars stayed in the same patterns night after night, a few of these bright points of light seemed to be moving

Figure 1: A view at sunset showing Venus in the western sky at the end of twilight.

Solar System Scoop

Stars Twinkle; Planets Don't

When you gaze up at the night sky, you may notice that some of the bright objects keep shining steadily. Those are the planets. The points of light that seem to twinkle are stars. But compared to the stars, the planets are miniscule, but are much, much closer. Because they are close, the light reflecting off their surfaces that reaches our eyes through the atmosphere travels along slightly different paths through the atmosphere, averaging out its effect, so the sum we see is a steady bright glow.

among the other stars. Those meandering objects that refused to stay fixed in space came to be known as "planets," from the Greek word for "wanderers."

The planets that shone in the ancient sky are the planets that form our solar system today. As a legacy of the early stargazers, their names come from the Greek and Roman mythology of more than two thousand years ago: Mercury, Venus, Earth, Mars, Jupiter, and Saturn.

In the next chapter, we'll see how the wandering of the planets among the stars was explained. For centuries, people thought that our Earth was the center and the other planets simply revolved around *us*. This Earth-centric view prevailed until the sixteenth century, when Copernicus figured out that it was the Sun, a star, and not the Earth that was at the center. And so the idea of our "solar system" was born.

Our Sun, the Center of Our Part of the Universe

Just what is this "Sun" that makes it the center of the solar system? For one thing, the Sun is huge. Nearly 200 Earths could span across its vast surface. Almost one million Earths could be packed inside it without even a mountain or two being compressed.

The Sun's humongous size comes with a lot of mass. It has hundreds of thousands of times more mass than our puny Earth, and hundreds of times more mass than all the planets together, gas giants and all. In this setting, size is power. With all that mass, the gravity of the single Sun dominates the eight planets. This central position has earned the Sun naming rights. The "solar" in "solar system," referring to our part of the universe, comes from the Latin word "sol," meaning "sun."

Basically, the Sun is a huge ball of gas. Its visible surface registers thousands of degrees hot and even that temperature is cool compared to the core, which is millions of degrees hot. The Sun is considered a typical star; that is, its size, mass, and temperature rank in the middle range of those properties among the billions and billions of stars lighting the sky. There are stars that are hotter and stars that are cooler; stars that are bigger and stars that are smaller. But lest the center of our solar system seem no more than average, there are many more smaller, cooler, less massive stars than larger, hotter, or more massive stars than our Sun. So compared with the bulk of the stars, the Sun is actually pretty hot, pretty, big, and pretty massive.

What are Planets?

As we said earlier, the planets were called the wanderers, as they rose, moved across the sky, and set the course of the night. Mercury, Venus, Mars, Jupiter, and Saturn were known since antiquity. So it came as a shock when an amateur astronomer named William Herschel discovered a new wandering object in 1781. At first, Herschel thought his find was a comet. But when it turned out to be orbiting the Sun beyond Saturn, all of his peers agreed it was a planet. There was less agreement on what to name the newly found object. After some skirmishing, it was named Uranus, from *Ouranos,* the god of the sky and the husband of Earth (Gaia). The tradition of naming celestial objects after mythological figures continues today.

The new planet named, the mathematicians of the Age of Enlightenment went to work analyzing detailed studies of the orbit of Uranus. Its orbit turned out to be more interesting than they thought. Instead of being completely regular, it had small deviations, which led to the idea that there must be another planet beyond it. It took more than sixty years, but in 1846, that new planet was found. Uranus, god of the sky, was joined by Neptune, god of the sea—the two outer planets bringing the total of planets to eight.

Smart Facts

(1 AU = astronomical unit, the distance of Earth from the Sun)

Mercury	0.4 AU
Venus	0.7 AU
Earth	1 AU
Mars	1.5 AU
Jupiter	5.2 AU
Saturn	9.5 AU
Uranus	19.1 AU
Neptune	30.1 AU

The twentieth century brought the discovery of yet another planet, or so people thought. In 1930, Clyde Tombaugh, a young astronomy assistant at the Lowell Observatory, found another object orbiting the Sun, even farther out in the solar system. Cold, distant, and mysterious, the object took the name Pluto, the god of the underworld. For seventy-five years, Pluto ranked as the ninth planet. But new discoveries, as we will see later on in this book, ultimately led to Pluto's demotion. As it turned out, the small, remote object had only 1/500 the mass of our Earth—too lightweight to deserve full-fledged planethood. And starting in the 1990s, a bunch of similar objects began to be found in Pluto's part of the solar system. Even more of a threat to the ninth planet's status, one of those objects seemed to be larger than Pluto. The question became which committee of the International Astronomical Union should name the newly identified objects. Would it be the committee that named planets and parts of planets? Or would it be the committee that named smaller things? Though it may seem hard to believe, the nature of what makes a planet had never been officially defined. For the first time in thousands of years, the scientists had to decide on a formal definition of "planet."

The showdown finally came at the General Assembly of the International Astronomical Union, in Prague, Czech Republic, in 2006. A planet was defined formally as an object that was not only round because of its own gravity, but also as something that "cleared its neighborhood." No one knew exactly what that last phrase meant (although the image of celestial objects named after gods clearing trash is intriguing), but it did effectively exclude Pluto, which had lots of similar objects also orbiting the Sun in its part of the solar system. Pluto did not clear its neighborhood and therefore, Pluto did not classify as a planet.

The newly discovered object larger than Pluto took the name Eris. Pluto, Eris, and the other objects that were massive enough to be round but not massive enough to clear their neighborhoods became "dwarf planets." Later on in this book, we'll be discussing the dwarf planets we now know about: Ceres, Pluto, Eris, Haumea, and Makemake. More of them will undoubtedly be discovered, perhaps hundreds of them eventually.

Objects in the solar system, shown in this set of NASA images. Shown are the eight planets and Pluto curving from lower center up toward top right, with a blazing sun. A comet and a couple of asteroids are also shown. The sizes are obviously not to scale.

Motions in Our Solar System

The objects in our solar system orbit the Sun; that is, they move around the Sun and come back to the same place, held to that path by the Sun's gravity. The movement around the Sun is called "revolution." At the same time, the orbits spin on their own axes, just as the Earth turns on its axis every twenty-four hours. This spinning movement is known as "rotation." If we measure the spinning by how it looks from outside the solar system, we call it a "sidereal" period, meaning by the stars.

The most massive objects in the solar system are the Sun and the eight planets, though as we established, the Sun is the center, with its immense size and powerful gravity. The planets, the dwarf planets, and a class of even smaller objects known as minor planets or "asteroids" all share the two properties of rotation and revolution; they all rotate on their axes as they revolve around the Sun.

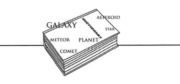

Solar System Words

When some new moving objects were discovered, beginning on New Year's Day 1801, it came as a surprise to their observers to see more than one moving in similar orbits. By 1810, four of these intriguing objects had been found. Sir William Herschel, who had discovered the planet Uranus about twenty years before the first of these novel objects appeared, called them asteroids, meaning star-like.

Instead of measuring distances in kilometers or miles (minuscule by the standards of space), it is often easier to consider distance compared with the average distance from the Earth to the Sun. This distance is known as one Astronomical Unit and abbreviated as 1 AU. (It covers 93 million miles, or about 150 million kilometers.) For example, since Mars averages 1.5 times farther away from the Sun than the Earth, the red planet's distance is 1.5 AU.

Eccentric Orbits

As we will see in the next chapter, the orbits of the objects revolving around the Sun are not quite round but "eccentric," meaning they slightly (or not so slightly) deviate from a circular path. Mars has the most eccentric orbit of the current list of planets. Pluto's orbit is so eccentric that it was long suspected of not being a real planet, even aside from its puny mass.

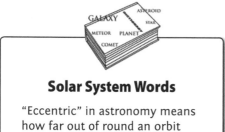

Solar System Words

"Eccentric" in astronomy means how far out of round an orbit is. The planets' orbits are pretty round, but not quite so. Their eccentricity refers to how far from round they depart. Comets, like Halley's Comet, have very eccentric orbits—their orbits look like really squashed circles.

Retrograde Motion

The ancients who labeled the planets wanderers, from the way they seemed to move among the stars, also noticed that most of the time the planets appear to move slightly faster than the stars, so that they moved ahead of the stars a little bit each night. But sometimes, they seem to move backward with respect to the stars, instead. This reversal is called "retrograde motion." In the next chapter, we'll see how interpreting retrograde motion led to our current understanding of the solar system.

The Solar System and the Zodiac

Like the planets, the Sun also moves across the sky at a different rate from the other stars. But since the Sun lights up the sky, we can't see the stars near the Sun (at least, not without a solar eclipse). Still, by looking at which constellations—patterns of stars—appear in the night sky, we can tell which constellation the Sun is moving through.

The twelve traditional constellations are Aries (the Ram), Taurus (the Bull), Gemini (the Twins),

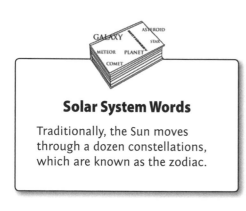

Solar System Words

Traditionally, the Sun moves through a dozen constellations, which are known as the zodiac.

Cancer (the Crab), Leo (the Lion), Virgo (the Harvest Goddess), Libra (the Scales), Scorpius (the Scorpion), Sagittarius (the Archer), Capricornus (the Sea Goat), Aquarius (the Water Carrier), and Pisces (the Fish). All these names date back thousands of years, but the stellar patterns are not in the same position relative to the Sun as they were 3,000 years ago. In 1930, the International Astronomical Union redefined the boundaries of the constellations so that the Sun now officially moves through a thirteenth constellation, Ophiuchus (the Serpent Holder), between Scorpio and Sagittarius. With changing boundaries, the Sun passes through parts of other constellations as well.

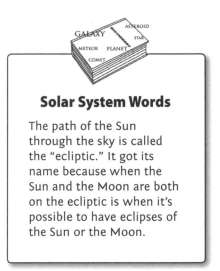

Solar System Words

The path of the Sun through the sky is called the "ecliptic." It got its name because when the Sun and the Moon are both on the ecliptic is when it's possible to have eclipses of the Sun or the Moon.

We can see the Sun move along the ecliptic because our Earth's orbit lies in a plane called the "ecliptic plane." Because the plane where the other planets' orbits lie is not very tilted with respect to the ecliptic plane, the other planets' orbits also lie in the zodiacal constellations.

Jay M. Pasachoff

An ancient astronomical clock, which shows the signs of the zodiacal constellations in the innermost ring. The ring moves in a way to show which constellation of the zodiac the Sun is in at any time during the year.

Terrestrials versus Giants

The planets in our solar system fall into two groups. Mercury, Venus, Earth, and Mars are the "terrestrial planets." They are relatively small and they have solid surfaces. Jupiter, Saturn, Uranus, and Neptune are "giant planets." They are much larger than the terrestrial planets and their surfaces are gaseous, not solid. Uranus and Neptune are sometimes called "ice giants," which highlights how far they are from the Sun and sets them apart from the much bigger gas giants, Jupiter and Saturn.

Starting with our home planet and moving outward, each of the planets has at least one object that orbits around it. Earth has one natural satellite, the Moon, joined since the mid-twentieth century by hundreds of artificial satellites launched into space by various countries. Giants Jupiter and Saturn both have dozens (and dozens) of satellites each.

NASA has sent spacecraft to all the planets, and in future chapters we'll discuss some of the findings. Each of the planets and each of the satellites is interesting in its own unique way.

Terrestrial Planets

Giant Planets

Planets in the solar system are shown to their approximate scale. The four terrestrial planets are much smaller than the four giant planets. Pluto, Eris, and the other dwarf planets are approximately half the size of Mercury..

Origin of Our Solar System

Clouds of gas and dust occur from time to time throughout the space between the stars. Sometimes it can be just a random motion that makes a particular cloud start coming together. The force of gravity among the various particles in it pulls it together faster and faster.

Smart Facts

Comparing the Planets' Sizes

Mercury	4,880 km diameter	38% of Earth's diameter
Venus	12,154 km diameter	95% of Earth's diameter
Earth	12,756 km diameter	
Mars	6,794 km diameter	53% of Earth's diameter
Jupiter	142,984 km diameter	11.2 times Earth's diameter
Saturn	120,536 km diameter	9.4 times Earth's diameter
Uranus	51,108 km diameter	4.0 times Earth's diameter
Neptune	49,528 km diameter	3.9 times Earth's diameter

In some cases, the collapse is triggered by an exploding star nearby. Many scientists have concluded that our own solar system's collapse was started by just such a supernova. They think this is how it happened because a radioactive form of aluminum survives in our solar system, meaning it must have been formed not long before the solar system was formed. And the same type of aluminum is formed in supernovae.

In any event, we know that our solar system was formed about 4.6 billion years ago. As the gas in the midst of the solar system converged, it drew energy out of gravity as it collapsed. Eventually, the gas became hot enough and dense enough that the hydrogen atoms in it started to fuse together to become helium. What Albert Einstein's famous formula $E=mc^2$ actually means is that the tiny bits of the original hydrogen that disappeared were transformed into a lot of energy. That is, those miniscule bits are the "m" in the equation, and are multiplied by the speed of light ("c") squared, and the speed of light is a very big number (roughly 186,282 miles per second, but we'll get to that later).

When the Sun began to shine from this fusion energy, it created a strong solar wind that drove a lot of particles out from around it. Some of the particles stuck together to make "planetesimals." In some cases, the planetesimals stuck together to make "protoplanets." Clearly, a lot of them stuck together to form Jupiter, whose huge mass attracted still more of the gas, dust, and particles that populated the early solar system.

In the infant years of the solar system, the objects had irregular orbits, bent by encounters (or collisions) with other objects. Eventually, most of the colliding objects were ejected out into space, leaving the solar system we have today (fortunately for us here on Earth).

Still, the solar system has far more inhabitants than the eight major planets and dwarf planets revolving around the Sun. In between Mars and Jupiter, thousands of planetesimals remain. We call them "minor planets" or "asteroids." Thousands—or millions—of the small objects populate space beyond Neptune as well. Pluto and Eris are now the largest of this collection, but as more objects are discovered, yet another one may outclass Pluto (or Eris) in size. These objects are called trans-Neptunian objects or Kuiper-belt objects.

Beyond the trans-Neptunian objects is an immense cloud of tiny, icy particles in the deep freeze of space. We call it the Oort Cloud, and it's where a lot of comets come from. Many objects found their way to the Oort Cloud after being ejected from the Kuiper belt. The Oort Cloud marks the end of our solar system, extending a substantial part of the way toward the nearest star.

Comets and Other Interplanetary Objects

Comets

They may lack the dazzle of stars, but comets are some of the most spectacular objects in the solar system. Balls of ice and dust, comets are known for their huge, long, and very beautiful tails.

The nuclei of comets may not seem impressive; in fact, they are dirty snowballs, only tens of kilometers across. But when these snowballs come in from the Kuiper belt or the Oort Cloud, the sunlight changes the ice directly into gas. Spacecraft flying close to Halley's Comet when it last visited our part of the solar system showed that the gas and dust it attracts spew out of the comet in jets. These jets eventually form a tail that can span so far across space that it's actually bigger than anything else in the solar system, despite being very diffuse.

Comets have very eccentric orbits; that is, their orbits are far out of round. As a result, we see comets only when they visit the inner part of the solar system. Most of the time, they are out in the far reaches of the solar system, unfortunately for us, way beyond the view of the human eye.

By far, the most familiar comet is Halley's Comet, which visits pretty regularly every 75 or 76 years and is fairly bright. The comet got its name from the English astronomer Edmond Halley, who predicted in 1715 that several sightings of bright comets were really a single comet that kept coming back time and again. The next time the glowing ball with its remarkable tail showed up in the sky, it became Halley's Comet. Halley's Comet was last near us in 1985–1986, so it won't be back big and bright until 2061. But don't be disheartened, there are brighter comets that make their way into our view from time to time. The brightest ones show up on short notice—only a year or so—so we can't even write about when you can expect to see the next Kuiper belt or Oort Cloud traveler. But if you read about it or hear about it, don't miss looking up at the sky for an amazing sight!

Halley's Comet during its 1985-86 appearance.

Meteoroids, Meteorites, and Meteors

The dust left behind in the tails of comets form tracks around the Sun, marking the comets' orbits. Every year at the same time, the Earth traverses those paths. When that happens, bits of the comet dust burn up in Earth's atmosphere, making meteor trails. The best time to catch a meteor trail is usually during the Perseid meteor shower that occurs each year on August 12, and is also visible a few days before and after that peak date. The meteors only appear to come from the direction of the constellation Perseus, since they are really comet dust in our solar system.

Most of the bits of comet dust from a meteor shower burn up in the Earth's atmosphere, tens of kilometers above our planet. Sometimes, though, we encounter a larger chunk of rock, which we call a "meteoroid." Some of them are extremely bright. The brightest ones—which can outshine Venus and cast their own shadows—are called "fireballs." Sometimes bits of meteoroids survive the fiery passage through our atmosphere. Once they land on the ground (or, in some cases, in cars, homes, or other buildings), they are known as meteorites. A few meteorites even come from the Moon and our neighboring planet, Mars.

CHAPTER 2

From Aristotle to Newton

In This Chapter

➤ Copernicus's revolutionary ideas

➤ Tycho Brahe, the noble observer

➤ Johannes Kepler and the harmonies of the planets

➤ Galileo sees the sky through a telescope

➤ Isaac Newton: Gravity and enlightenment

The Earth at the Center

In ancient and medieval times, and even into the Renaissance, Western thinking was dominated by the ideas of Aristotle (384 BC–322 BC), the renowned Greek philosopher and scientist. Aristotle's view of the universe was based on the four elements of Earth, Air, Fire, and Water, to which he added a fifth element: Aether (or *quintessence* from "quint" meaning "fifth"), to refer to an immutable, non-material (that is, non-earthly) substance that made up the celestial bodies of stars and planets. And according to Aristotelian physics, which held for more than one thousand years, all of those gleaming celestial bodies revolved around *us*. Earth stood at the center of the universe.

Without our modern light pollution, it was easy to see the stars on a clear night, tracing their paths as they rose in the eastern sky, traveled upward, and finally set in the west. Venus stood out for its brightness, but there was another star that attracted stargazers because it seemed to be fixed in space, while the other stars slowly circled around it. That star became known as Polaris, the North Star. Polaris is not all that bright; forty-eight stars outshine it. But Polaris earned its fame by its intriguing place in the sky. In the universe, location really can make a big difference.

With the heavenly objects a nightly attraction, it became obvious to the ancients that not all of the luminous objects rose and set precisely with the stars. That was especially true of the brightest object of all, the Moon. This brilliant and mysterious object had its own peculiar schedule. It rose nearly an hour later each night, surrounded by a different backdrop of stars. Even more unusual, its shape changed, from a slender crescent to a full, bright circle and back again. We now identify this pattern as the moon's phases.

The ancients also recognized that the dazzling objects moving across the sky—the wanderers, or planets—were different from the stars. They could also pick out certain features that distinguished one from another, which may have influenced the Romans, whose names for the planets we still use today. Venus, the brightest of the planets, was named for the goddess of love and beauty. Giant Jupiter, also dazzlingly bright, carries the name of the king of the gods. Reddish-tinged Mars bears the name of the god of war, and Mercury, speeding faster than all the other planets, has the name of the fleet-footed messenger of the gods. Saturn, beyond Jupiter and fainter than the other planets, has the name of the god of agriculture and father of Jupiter.

Moving Backward

Most of the time, the planets were pretty predictable in their nightly meandering. Like the stars, they rose in the east and traveled west. Some nights, though the wanderers seemed to pause and reverse their path, moving backward with respect to the stars. The name for this backward motion is "retrograde" motion, in contrast to "prograde," the forward motion.

In the great city of Alexandria (now part of Egypt), a center of learning in the ancient world, the Greek astronomer Claudius Ptolemy (AD 90 c.–AD 168) was intrigued by the planets' retrograde motion. In the second century AD, Ptolemy decided he found the solution to the curious backward movement. Ptolemy's theory firmly places the Earth at the center of the universe. The central theme is that the Earth is at the center and the Moon, and the planets move around it in circles that keep getting bigger and bigger. Imagine that the planets actually move on small circles with centers that are on the big circles. For part of the time, in that scenario, they would be traveling backward faster than the centers of the circles are traveling forward. Stargazers on Earth would be watching the planets move backward.

The small circles are known as "epicycles," and the large circles are "deferents." For more than a thousand years, Ptolemy's concept of epicycles moving on deferents dominated scientific thinking on planetary motions.

Centuries before the telescope was invented (in 1609), observing the planets was fascinating, no doubt, but not very accurate. With the positions of the planets based on how they appeared along poles with respect to the stars, Ptolemy's theory actually worked pretty well with the knowledge they had at the time. It was also easy to tweak to make it work even better. The epicycles could have more epicycles on them; that is, the small circles could have even smaller circles, and the smaller circles could have smaller circles…and on and on, until the theory was finally disproved.

Copernicus and the Revolution(s) in Thinking

In the early sixteenth century, a young cleric named Nicolaus Copernicus (1473–1543) was sent from his home in what is now Poland to further his education in Italy as a medical student. He eventually returned home to work as assistant to his uncle, a powerful bishop. Copernicus had an admirable education, immersed in science, mathematics, and humanities. He had ideas about money management that would have made him a good economist by today's standards. But it was astronomy that was Copernicus's driving interest, and he is most famous as an astronomer. He was also a master mathematician, which he applied to his astronomical theories.

Copernicus had an idea that was very radical in an age still dominated by Aristotle. He thought it would be simpler to understand the structure of the universe if its center was the Sun, not the Earth. With the Sun at the center, the planets' retrograde motion would happen even if all the planets simply circled around the Sun.

Smart Facts

The backward motion would actually be an optical illusion that takes place when the Earth passes another planet. For example, if you are standing still next to someone and you start moving forward first, it might look for an instant as if the other person is moving backward.

Copernicus was savvy enough to realize that his ideas could prove controversial. They became known throughout Europe, but Copernicus made no move to publish his work. Then, in 1539, a young mathematician named Georg Joachim Rheticus became Copernicus's pupil. Finally, he persuaded Copernicus to publish his work in book form, even getting permission to take the manuscript to the publisher. The title was to be *De revolutionibus,* or *On the Revolutions.* That title was eventually changed (slightly) after a cleric who supervised the final stages of the manuscript's publication raised the fear that Copernicus could find himself in the midst of a storm for bringing up ideas that could be seen as challenging Scripture. (For example, Joshua commands the Sun to stand still.) Ultimately, it was Galileo who was embroiled in a dispute with the Church over the idea of a heliocentric versus geocentric (Sun-centered versus Earth-centered) universe, as we will see later in this chapter.

Solar System Scoop

The change to the manuscript title was that the words *Orbium Coelestium* were added to *De revolutionibus* on the first page of type, translating to *On the Revolutions of the Celestial Spheres.* In that way, the words could be interpreted to mean that the "revolutions" referred only to celestial bodies rather than scientific truth.

Copernicus's diagram showing the Sun at the center of what we call the Solar System. This diagram is from the first edition of *De Revolutionibus Orbium Celestium (On the Revolutions of the Celestial Orbs),* published in 1543.

Inside Copernicus's reluctantly published book is probably the most famous diagram in the history of science. At the center of the illustration is the Sun, with a series of circles around it. Copernicus's mathematical genius served him well as an astronomer. He knew the speeds of the planets revolving around the Sun, taken from how fast they seemed to travel in the sky. With that knowledge, he could put them in the correct order: Mercury, Venus, Earth, Mars, Jupiter, and Saturn. The planets orbited the Sun between the circles illustrated in the diagram.

Copernicus was able to calculate how fast the planets were traveling by how they appeared to move in the sky. His calculations were based on how long the planets took to revolve around the Sun if we were gazing down on them from high overhead, instead of from a planet that was moving in its own orbit around the Sun. Contemporary astronomers use simple algebra to chart the equation discovered by Copernicus. The formula allows time for our earthly home to catch up with the outer planets, which have pulled ahead during our solar year.

Solar System Scoop

In sixteenth-century England, mathematician and scientist Leonard Digges wrote a book in England about the Earth and the universe. His son Thomas is credited with being the first person to describe Copernicus's ideas in English, in 1576 in an appendix to a new edition of his father's work. The appendix includes the first English-language version of Copernicus's famous diagram.

The first telescope was still decades away. Copernicus could pride himself on the fact that his model of the solar system did away with the need for epicycles, except very minor ones. It is still clear, though, from the diagram that Copernicus's model, heliocentric though it was, still relied on circles, which were considered fundamental. Copernicus occupies an interesting position in the evolution of the science of astronomy. His work marked the beginning of modern astronomy, but he did not completely break free from the ancient teachings.

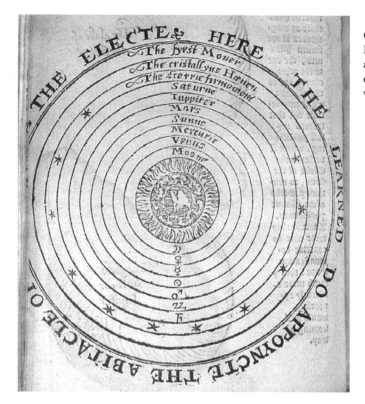

Copernicus's first diagram in English, from Thomas Digges's appendix to "A prognostication everlasting," a book by his father that was first published in 1556.

The Start of the Modern Observatory

Being in charge of an island may seem an unusual path toward a new science, but Tycho Brahe (1546–1601) was pretty unusual for a scientist (for one thing, he wore a metal prosthetic nose after losing part of his nose in a dual). A Danish aristocrat by birth, he was given control of the island Hven off the coast of Copenhagen and collected rent from the people living there. By the standards of any era, Tycho was wildly rich. He also seemed to have had the ego of a contemporary mogul, who wants everything he has to be bigger and better than everyone else's. In Tycho's case, beginning in 1575, he used his vast wealth to build an observatory with bigger and better instruments. The instruments were still primitive, but they still surpassed other instruments of the day in their precision for measuring the positions of the planets and stars.

Tycho's home was a castle. It was also something of a forerunner to our modern research institutes, where visiting scientists had plenty of space and equipment to work. Being a guest meant being able to work with what was then state-of-the-art scientific equipment, and enjoying elaborate dinners, complete with a jester for entertainment.

Tycho invested far more energy in his scientific endeavors than in his duties to the Danish monarchy. Not a good political move. When a new king and queen took the throne, Tycho lost his position. He decided to move to Prague, where he became court mathematician to the Holy Roman Emperor Rudolf. He even took his printing presses and pages that had been printed, but he no longer had his observatory to the stars.

Tycho is definitely credited with improving the accuracy of scientific observations. Some other achievements are more dubious. Tycho wanted to outshine Copernicus by creating his own model of the solar system. It was an interesting cross between the theories of Ptolemy and Copernicus. In Tycho's fusion model, the Sun and the Moon both revolve around the Earth, while the other planets revolve around the Sun. Today, it is only mentioned (if at all) as one of many curious theories of the universe.

Tycho's main contribution to modern astronomy is a wealth of high-quality observations of Mars moving among the stars. Even that is somewhat indirect. When Tycho left Denmark, he invited a young mathematician named Johannes Kepler to join him in Prague. Tycho's legacy comes to us mainly though Kepler's work.

From the Mystery of the Cosmos to the New Astronomy

As a six-year-old child, Johannes Kepler (1571–1630) saw the Great Comet of 1577, and two years later, he witnessed a lunar eclipse. Those images stayed with him and inspired a lifelong love of astronomy. Kepler was teaching mathematics in Graz, Austria, but his mind seemed to be filled with astronomical images. He realized that there are five regular polygons, or forms composed of equally shaped sides, and with some experiments, he found that if these solid forms were nested and separated by spheres, they could correspond to the six known planets. In 1596, Kepler published his book on the topic, *Mysterium Cosmographicum,* or *Mystery of the Cosmos.* The book contains a very beautiful diagram in which an octahedron, icosahedron, dodecahedron, tetrahedron, and cube are nested, with their eight, twelve, twenty, four, and six sides, respectively, meant to illustrate the spacing of the six planets Mercury, Venus, Earth, Mars, Jupiter, and Saturn. Aesthetically, the diagram is stunning. Scientifically, it is totally wrong.

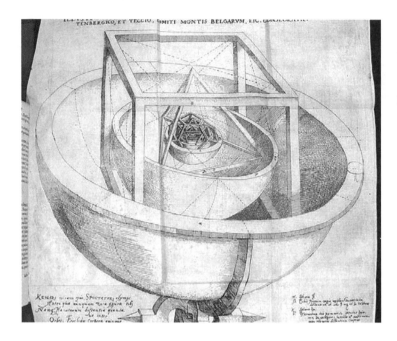

Kepler's fold-out diagram from his *Mysterium Cosmographicum (Mystery of the Cosmos)*, which was published in 1596.

Despite its flaws, Kepler's work earned him a reputation as a skilled and dedicated astronomer. He was even granted an interview with the Great Man, Tycho Brahe, and in 1600, he became one of Tycho's assistants, making calculations based on Tycho's observations of Mars. Typically, Tycho was not generous in sharing his data, but he was impressed enough by Kepler's ideas to grant him more access to his trove of astronomical data.

Tycho's unfortunate death is well known. As a guest at a banquet in Prague, he drank heavily, but according to legend, etiquette kept him from leaving the table. Ultimately, he was said to have died of a ruptured bladder, with his last agonizing hours described in detail in Kepler's diary. (There was some recent suspicion that Tycho died of mercury poisoning, with Kepler listed as one of the possible suspects. But that theory has since been dismissed, and a ruptured bladder seems the most probable cause.)

Every era has its own red tape and bureaucracy, and it took awhile before Kepler could get Tycho's data. Because Tycho had married a commoner, his property could not be inherited by his family, and his scientific data was among the few valuable things they could get. Eventually, Kepler got access to Tycho's data on Mars, and he tried to fit it to his calculations.

What Kepler got from Tycho were his observations of the position of Mars in the sky surrounded by stars. Using that data, he was determined to calculate the red planet's real motion around the Sun. First, he attempted to fit a circle or an egg-shaped orbit, but as many times as he tried, it was never an accurate match to the observations. Finally, he got the idea of using an ellipse, a geometrical figure that looks sort of like a squashed circle. That was Kepler's "Eureka!" moment—this time, it fit!

Kepler's discovery is one of the highlights of the history of astronomy. As it turned it, it applies not only to Mars and the other planets in our solar system, but to the hundreds of planets we have been discovering orbiting other stars.

In 1609, Kepler published *Astronomia Nova* (*The New Astronomy*), which included the first two of his three laws of planetary motion:

➤ Kepler's first law of planetary motion: Every planet orbits the Sun in an ellipse, with the Sun at one focus.

➤ Kepler's second law of planetary motion: A line joining a planet and the Sun sweeps out equal areas at equal time intervals.

Solar System Words

An ellipse is a closed curve that has two key points called the foci. To draw an ellipse, put two dots on a piece of paper. Those are the foci. Then, take a piece of string longer than the distance between the two foci and tape them down right on the foci. In the next step, put a pencil or pen in the string and pull it taut, then move it gently from side to side. The curve you see is one side of the ellipse. From that, you can see how to move the pencil or pen to draw the other side. This experiment illustrates one of the defining points of an ellipse: the sum of the distances from a point on the ellipse to each focus stays constant.

Moving in Harmony

Music pervaded Kepler's ideas of the planets, and he was forever searching for harmonies in the movements of the planets. To Kepler, the speeds of the planets were linked with musical notes. The higher notes corresponded to the fast-moving inner planets Mercury and Venus. *Harmonice Mundi* (*The Harmony of the World*) appeared in 1619. Now recognized as Kepler's foremost work, the book contains his third law of planetary motion *and* musical staffs with notes, an interesting combination. The third law goes like this:

➤ Kepler's third law of planetary motion: The square of the orbital period of a planet's orbit around the Sun is directly proportional to the cube of the size of its orbit. Because the period of an orbit can be measured through observation, this law allows scientists to infer the orbits' relative sizes.

Smart Facts

Distances from the Sun and Periods of the Planets known in Kepler's Time

Mercury	0.387 AU	0.240 year
Venus	0.723 AU	0.615 year
Earth	1 AU	1 year
Mars	1.523 AU	1.881 years
Jupiter	5.203 AU	11.857 years
Saturn	9.537 AU	29.424 years

As we discussed in Chapter 1, astronomers usually measure the distances between planets by using the Astronomical Unit (AU). Now we add the periods of the planets known in the seventeenth century.

If you square the second column (the periods) and cube the first column (the distances), you can see that the two figures are pretty much equal. Any discrepancies result from approximation, not actual differences. (Extending the number of decimal places would more accurately capture the influences of other planets.)

Turning the Telescope to the Sky

The year 2009 was a stellar year for astronomy. First, the International Astronomical Union declared it to be the International Year of Astronomy. Then UNESCO and finally, the United Nations General Assembly did the same. What was the celebration about? Exactly four hundred years ago, in 1609, Galileo Galilei turned the first telescope skyward. (Kepler's fans were definitely *not* pleased by the commemoration. From their view, the 1609 publication of Kepler's *Astronomia Nova* should hold equal importance in the history of astronomy).

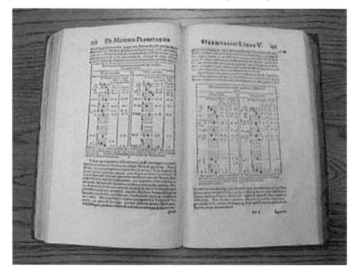

From Kepler's *Harmonice Mundi (The Harmony of the World)*, which was published in 1619. There are musical notes to show the different pitches from the different planets.

As a professor in Padua, Italy, Galileo heard of a Dutch invention that could make distant objects appear closer than they actually were. Galileo had never seen this talked-about instrument, but he used his knowledge and ingenuity to figure out the type of lenses it would have and created his own. Padua was close to Venice, which was a thriving seaport. Galileo presented his new device to the Venetian nobility as a tool for commercial and military ventures. With this novel gadget, ships could be seen far out in the sea. The noblemen were impressed, and Galileo's invention was a tremendous success.

Galileo was driven by scientific curiosity. If his device could make ships at sea seem closer, then why not the glowing objects that light up the sky? Of course, innovative as it was, Galileo's telescope was still a primitive piece of technology, and it was not very easy to use. The field of vision was very narrow, and it was hard to focus. But despite its obvious limitations, Galileo aimed his telescope upward and found he could see part of the Moon. And through his lenses, he could see that the Moon had a lot of formations. With his training in drawing in Renaissance Italy, Galileo realized that he was looking at areas of light and shadow, and he recognized what they were. He had discovered the mountains and craters that dot the Moon. He could even calculate their height from the length of the shadows they cast and the way they varied from the changing light of the Sun.

Smart Facts

Galileo was actually not the first to aim a telescope at the Moon. Thomas Harriott in England had done the same thing several months earlier, but he only recorded what he saw in rough sketches and scribbles. Harriott made other discoveries, but he never cared about publishing his work (he seemed to be indifferent to glory), and it was not until much later that his work became known.

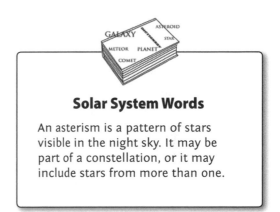

Solar System Words

An asterism is a pattern of stars visible in the night sky. It may be part of a constellation, or it may include stars from more than one.

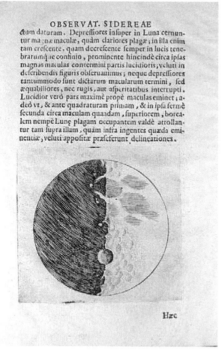

A page from Galileo's *Sidereus Nuncius (The Starry Messenger),* which was published in 1610, showing an engraving of the Moon. The book written in Latin, included extensive coverage of the relative motion of the four major moons of Jupiter.

Galileo was enterprising in presenting his telescope to the Venetian nobility, and he was definitely interested in publishing his discoveries. He began his writing in late 1609 in Padua. One discovery led to another as he pointed his telescope at other objects beyond the Moon. He could see that the Milky Way was home to numerous stars and saw many stars in some of the asterisms in the night sky. He also observed four "stars" near Jupiter that moved with the great planet and changed position each night. That sighting turned out to be the discovery of the major moons of Jupiter, now called the Galilean satellites. Galileo carefully chronicled his discoveries, which were published in a slim volume entitled *Sidereus Nuncius (The Starry Messenger)* in 1610.

Galileo's ambition and innovation earned him much admiration and fame. But not all of Renaissance Europe was as welcoming of new ideas and discoveries as cosmopolitan Venice. Using his discoveries to secure a higher status, higher paying position in Florence proved not to be a great career move for the brilliant scientist and inventor. Closer to Rome, the seat of Papal authority, Florence was far less liberal than the Venetian Republic. It was not the ideal place for a scientist with an exploring mind whose work was built on a heliocentric view of the universe.

The idea that the Sun, not the Earth, was the center of the universe challenged Aristotle's geocentric model that dominated Church thinking. In 1616, Galileo was summoned before the Inquisition and ordered not to teach Copernicanism. He pledged to obey and at least publicly, he stayed quiet on the subject for many years. According to one story, Galileo's teacher was burned at the stake for heresy, which might have inspired his silence. Then in 1632, Galileo published his *Dialogo* (Dialogue), which had three characters discussing the heliocentric and geocentric systems. He had official permission to publish the book, but the defender of the Earth-centered system often came off looking foolish, and it was very clear which astronomical system the author preferred. Galileo was put on trial and first sentenced to imprisonment, which was commuted to house arrest the next day. He remained under house arrest for the rest of his life.

Remember, Galileo was tried by the Inquisition, and torture was a real possibility (or probability) if he refused to recant his "heretical" view that the Earth moved around the Sun. According to legend, Galileo recanted before his Inquisitors, but muttered under his breath, "And yet it moves." There is virtually no real support for the apocryphal story, but it is easy to see why it remains a popular legend so many centuries later.

Newton's Laws

A symbol of the Age of Enlightenment, Isaac Newton (1643–1727) was unquestionably one of the greatest physicists of all time. In one simple phrase: Newton discovered the law of gravity.

Unlike Galileo's supposed act of rebellion, the famous story of how Newton discovered gravity is mostly true. The plague forced the English universities to close and the young student Newton was sent home from Cambridge University to Woolsthorpe in Lincolnshire, England. While he was there, he saw an apple fall from a tree (the dubious part is that the apple fell on his head), and he realized that the force that controlled the apple's descent from the tree was the same force controlling the motion of the Moon in the sky.

Newton eventually returned to Cambridge's Trinity College as a faculty member. At the same time, a group of scientists in London met in a coffeehouse to form a society that would later become the Royal Society. The group had heard that there was a genius mathematician named Isaac Newton who could help them solve a major scientific question. Young Edmond Halley was sent on the long and arduous journey by stagecoach from London to Cambridge (an hour today, though today they'd just have an online videoconference) to find Newton and see if he really was up to the task.

Halley asked Newton what would be the shape of an orbit if there was a force that fell off with the square of the distance. An ellipse, Newton responded. Intrigued and excited, Halley asked Newton if he had proved this idea. Newton told him he had the proof in some papers,

Smart Facts

Newton's classic book might not have been published without Halley. Halley paid for the printing, championed the work, and even wrote a preface.

Smart Facts

Newton's first law of motion: A body in motion tends to remain in motion, and a body at rest tends to remain at rest. (This one is probably Newton's best-known law, the law of inertia.)

Newton's second law of motion: force equals mass times acceleration (the modern version of Newton's law).

Newton's third law of motion: For every action, there is an equal and opposite reaction.

Halley's visit inspired Newton to document some of his mathematical conclusions. His most famous book, the *Principia Mathematica,* or the *Mathematical Principles of Natural Philosophy* (philosophy then included the sciences), was published in 1687 in Latin. Newton was a don at Trinity College at the time.

The *Principia* presents basic laws that govern the physical world in a single volume. The book includes Newton's law demonstrating how gravity diminishes by the square of the distance, along with his proof of Kepler's laws of planetary motion, and most important, Newton's laws of motion, clearly defined in the text as "laws." (Kepler's laws are written as merely a small part of Newton's text.)

Newton's apple tree in Woolsthorpe eventually had to be cut down due to old age. As befitting a tree that left such as powerful legacy, it was not simply chopped down and hauled away. Cuttings were taken from the tree and bits of wood were distributed, though if all the claims of cuttings and pieces of wood from Newton's apple tree were really true, the tree would have had to have been immense!

Newton became Sir Isaac Newton when he was knighted by Queen Anne in 1705. To most of us, the knighthood seems obvious, in view of his scientific genius. In reality, it was more likely because of his work as Master of the Royal Mint. There was no prize for physics in Newton's day.

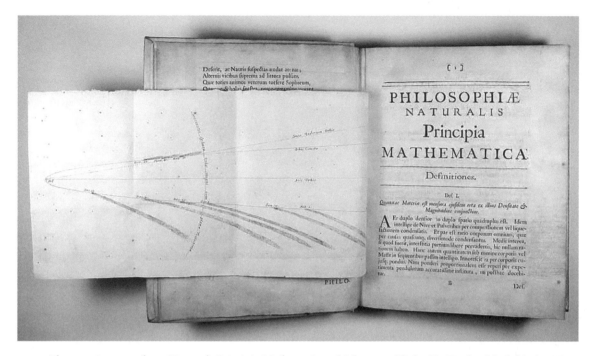

The opening page from Newton's *Principia Mathematica*, which was published in England in Latin in 1687. A fold-out chart of the path of Halley's Comet was included.

CHAPTER 3

From the Age of Enlightenment to the Space Age

Newton's Heirs

Isaac Newton's *Principia Mathematica* marked a tremendous breakthrough in the understanding of gravity. Not only was it invaluable to a new generation of scientists, but it also reinforced Kepler's place in the history of astronomy. Using his law of gravity, Newton was able to explain the formulas that Kepler arrived at decades earlier (in 1609 and 1619), when he analyzed Tycho Brahe's painstakingly measured but pre-telescopic data. Kepler's laws, the product of trial and error, were now grounded in firm scientific theory.

Watching the stunning sight of a comet as a small child led Kepler to a life in astronomy, as fellow scientists Newton and Halley both watched comets in 1680 and 1682. Newton published *Principia* in that same decade, in 1687, but soon after, that he left science for government, becoming Master of the Mint in 1696 and ultimately becoming Sir Isaac. Edmond Halley (1656–1742) remained fascinated by comets.

Despite his interest in comets, most of Halley's scientific work in the late seventeenth-century was more earthbound, which may be why it was not until 1715 that Halley realized the significance of his own and others' observations of comets. Gathering all the known information on comets, Halley used Newton's laws to determine their orbits. Some of the comets seemed to have orbits that were unusually similar. In fact, the data revealed a pair with intervals that were almost, though not quite, the same. He suspected that Jupiter's gravity might have slightly altered the orbit, meaning it could be the same comet returning back into view. With that idea, Halley predicted that the comet would return once again in 1758.

Halley's death predated the comet's proposed return by sixteen years. During that time, other scientists analyzed and refined Halley's predictions for when the comet might reappear. When 1758 arrived, they kept watching the skies for the greatly anticipated event, but there was no blaze of light and no gossamer tail. The year was almost over when finally, on Christmas Day, a farmer and amateur astronomer saw the comet streaking across the night sky. With Halley's prediction proved true, the most famous comet in history was named after the scientist who foresaw its return.

Since antiquity, philosophers, scientists, musicians, poets, and countless stargazers of all types have connected the celestial objects with music. Kepler might have been happy to know that William Herschel (1738–1822) was a musician and composer as well as an avid astronomer. Originally from Hanover, Germany, Herschel moved to Bath, England, later inviting his sister Caroline Herschel to join him as his assistant. William's interest in astronomy grew from his love of music, and in England, he began building telescopes and gazing up at the skies.

A lot of William Herschel's observations focused on "double stars"; that is, pairs of stars that appear very close together. One day in 1781, he saw what he initially thought was a star or a comet. As we discussed earlier, that discovery was the planet Uranus.

In 1783, William gave Caroline a telescope so she could "sweep" the skies in search of yet undiscovered objects that might come into view on a clear night. The enterprising Caroline discovered eight comets. The story of William and Caroline Herschel highlights the attraction gazing the heavens has held for millennia. Brother and sister both took up astronomy as a hobby, and the more they explored the skies, the more absorbed they became.

Solar System Scoop

Caroline and William Herschel both received stipends from King George III, which made Caroline (1750-1848) the first woman to be paid for work in astronomy. Later, she became the first woman to receive a gold medal from the Royal Astronomical Society. Comet 35P/Herschel-Rigollet is named for Caroline Herschel, who discovered it on December 21, 1788.

The 19th Century: An Age of Many Discoveries

The nineteenth century began very auspiciously for astronomy. On New Year's Day of the new century, the Italian astronomer Giuseppe Piazzi saw a novel object traveling across the night sky. He noticed that as it moved through the stars, it traveled faster than Mars, but slower than Jupiter. Piazzi named the new object Ceres, for the Roman goddess of agriculture. At first, the object was hailed as a new planet, in orbit between Mars and Jupiter. Why not another planet? It was not even twenty years since William Herschel discovered Uranus.

And it was not even another year before a second object was found orbiting the space between Mars and Jupiter. This new object was called Pallas, after Pallas Athena, an alternate name for Athena, goddess of wisdom. Things were getting very curious in the sky. No one had expected to find two objects in a similar orbit. It soon became even more curious (and crowded) when two other such objects were discovered within a few years. By 1807, Ceres and Pallas were joined by Juno and Vesta. These last two objects were named, respectively, for the Roman queen of the gods and the goddess of the hearth.

Solar System Words

These newly discovered objects retained their status as planets for half the 19th century. It would be 38 years before any more were discovered. But though not officially demoted, as early as 1802, these smaller objects were called "asteroids," for "starlike," a term that comes to us from William Herschel. Actually, "planetoids" would be more accurate for objects that are within our solar system and are definitely not stars. The term asteroids stuck; the objects' (precarious) status as planets officially ended in the 1850s.

The late nineteenth century was a time of tremendous breakthroughs in our knowledge of the spectrum of the Sun and stars. These discoveries allowed beams of light to be broken down into their constituent colors, and ultimately led to the development of astrophysics. Recognizing the great potential for discovery of collaborative projects (which drive contemporary astronomy), many observatories, especially in Europe, joined together to create a detailed map of the sky. Outstanding discoveries would follow, but at the time there was one casualty of the excitement over the spectrum and charting the positions of stars. New discoveries within the solar system came to a virtual standstill.

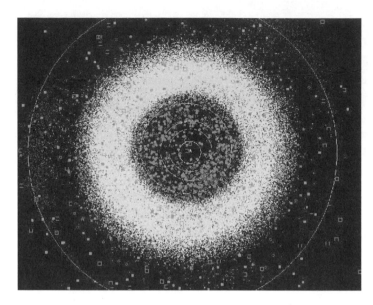

Tens of thousands of asteroids are shown by the bright dots. The outer circle is the orbit of Jupiter. Inner circles, from inside out, are Mercury, Venus, Earth, and Mars.

As far as *potential* discoveries in all parts of the universe, the nineteenth century, which began with the discovery of Ceres and her sister asteroids, ended with a new direction in the construction of astronomical observatories. Traditionally, observatories were built in cities in proximity to the universities or other institutions sponsoring them. By the late nineteenth century, observatories were beginning to spring up on mountaintops; there, far from city lights and industrial pollution, they could produce far superior observations.

The Solar System versus the Lure of Distant Galaxies

With the ability to scan far out in space, the solar system lost its attraction for many astronomers. Why stay in your own neighborhood when there's a whole universe filled with glittering objects waiting to be discovered by a powerful telescope? Mount Wilson Observatory in California was home to two of the biggest, most powerful telescopes of the early twentieth century: the 60-inch Hale telescope built in 1908, and the 100-inch Hooker telescope (named for John D. Hooker, who supplied the funds for its massive reflector), which held the distinction of being the world's largest telescope for three decades, from the time it was built in 1917 until 1948. (The giant telescope was eventually surpassed in size by the 200-inch telescope on top of Palomar Mountain, southeast of Mount Wilson.)

Beyond the Milky Way

Mount Wilson's most famous astronomer in the first half of the twentieth century was Edwin Hubble (1889–1953), who arrived a few years after the huge Hooker telescope was built. Using the cutting-edge telescope, Hubble proved unconditionally that there were other galaxies beyond the Milky Way. Though the controversy was nothing like the zealous dispute over the geocentric versus heliocentric view of the universe, Hubble's findings were initially challenged by astronomers who insisted the Milky Way galaxy was the entire universe. As we know, Hubble was right. He calculated the distances to several remote galaxies, and even more revolutionary, his measurements showed that the universe is expanding. Also part of his legacy, the Hubble sequence, his system for classifying galaxies, is still widely used by used by professional and amateur astronomers. To most of the world, Hubble's name is most familiar for the Hubble Space Telescope, transmitting stunning images and a countless amount of data since 1990.

Back in Our Solar System

Hubble exemplifies the generation of astronomers lured by the idea of new discoveries outside our solar system. But if our part of the universe seemed less exciting to most, there were some astronomers who felt the objects revolving around our Sun were still worthy of exploration. One of those was Gerard Kuiper (1905–1973) of the University of Chicago. Kuiper established an observatory in Texas (originally run by the University of Chicago, but now the McDonald Observatory of the University of Texas). He discovered two of the moons of the outer planets: Uranus's moon, Miranda, and Neptune's moon, Nereid. Nereid was a sea nymph who served Neptune, which fits in with the mythological naming of planetary objects. In a dramatic departure from that tradition, Kuiper named his discovery Miranda, after the heroine of Shakespeare's *The Tempest*. Kuiper's interest in our own solar system earned him lasting recognition in the Kuiper belt objects, that ever-growing collection of small but important objects beyond Neptune.

There were two other very significant discoveries in the mid-twentieth century just before the dawn of the Space Age. Fred Whipple at Harvard College Observatory concentrated on comets and concluded that they were essentially dirty snowballs (though with long, beautiful tails). That analysis has since been scientifically validated and is central to what we now know of comets. In the Netherlands, Jan Oort theorized that there is a vast, distant cloud of icy embryonic comets orbiting the Sun. It's now known as the Oort Cloud.

The dark silhouette reveals Halley's Comet as the European Space Agency's Giotto mission flew by it in 1986, showing jets of gas and dust being expelled.

Launching the Space Age

When Disneyland opened in July 1955, one of its main attractions was Tomorrowland. With an eye to the future of space travel (among other innovations), the visionary Walt Disney declared that "Tomorrow can be a wonderful age." Two years later, the Soviet Union astounded the world with the launching of Sputnik on October 4, 1957.

The world was caught up in the Space Age.

Officially, 1957–1958 was labeled the International Geophysical Year, as scientists from countries throughout the globe teamed up for an ambitious 18-month project spanning all areas of Earth science. The international collaboration aimed to examine the Earth's magnetic field, the auroras, and other elements of the Earth from the North to the South poles. The Soviets made it known that they were planning to launch a satellite into space, but it still took the world by surprise when Sputnik was launched into orbit around the Earth. Long the province of science fiction, the idea of rockets in space was now a reality. Sputnik's "beeps" transmitted back to our planet were picked up by Ham radio enthusiasts, who excitedly shared the news that it was real. They were joined by "Moonwatch" teams of amateur stargazers around the world, armed with small telescopes that could possibly zero in on the light reflected back by Sputnik as it traveled across the sky.

That event, still remarkable for its time decades later, is now known as Sputnik 1. A month later, on November 3, 1957, the Soviets once again stunned the world with the launch of Sputnik 2, carrying the dog Laika.

The Soviets launched the first satellite, the Space Age—and the Space Race. And the United States seemed awfully far behind. In 1955, the U.S. began Project Vanguard, planning to launch a satellite into space for the International Geophysical Year. It was expected to be the first artificial satellite sent into orbit. Not only did the Soviets beat the U.S. as number one into space, but two months after the stellar success of Sputnik, the first Vanguard rocket fell back and exploded on the launch pad. The effort was a hasty attempt to catch up with the Soviet Union. Only, instead of showing the world that the U.S. was a force to be reckoned with in the race for space, the result was a very public failure. A *Time* magazine article called it "kaputnik."

It's not surprising that the humiliating failure should lead to a shake-up of leadership. The Army lost its claim to the first American rocket launch to the Navy. The project was placed under the direction of Werner von Braun, a German scientist who was brought to the U.S. after World War II. If 1957 ended ignominiously for the American space effort, 1958 started off with a very positive bang. On January 31, Explorer 1 was successfully launched into space. The rockets were designed to conduct experiments that could be monitored back on Earth. Explorer I was equipped to study the Earth's magnetic field in an experiment devised by James Van Allen of Iowa State University. It succeeded magnificently! Its success is evident in the name given the discovery that eventually came from the project. The belts of charged particles surrounding the Earth and other planets are now known as the Van Allen belts.

The Vanguard rocket exploding as it was launched at Cape Canaveral, Florida, on December 6, 1957. This was about two months after the launch of Sputnik-1 and about one month after the launch of Sputnik-2, which carried the dog Laika into space.

Solar System Scoop

The first Vanguard rocket was unquestionably a huge failure, but a second Vanguard, launched a month after Explorer 1, succeeded in getting a satellite into space. In fact, the small, successful Vanguard has outlasted other early satellites—all burned up in Earth's atmosphere—making it the longest-lasting manmade space object.

Exploring the Solar System

Jules Verne wrote *From the Earth to the Moon* in 1865. The Moon and the other planets, especially our red neighbor Mars, had always seemed tantalizing targets for space exploration. With the successful launching of rockets in space, those objects in our solar system were starting to seem closer, and the prospect of sending rockets to the Moon and the planets began to seem more like reality. Later in this book, we'll go over the discoveries of the planets and their (many) moons in detail. For now, we will summarize a few major missions made possible by the onset of the Space Age.

To the Moon

Begun in the 1960s, the lunar program included robotic and manned missions. Once again, the Soviets were the first, electrifying the world in 1966 by flying a rocket around the Moon and capturing the first images of its dark side. The U.S. was still playing catch-up. That same year, though, the U.S. advanced in the Moon race, first sending Lunar Orbiters (1966–1967) and then Surveyor landers (1966–1968) to the Moon. The Moon's rocky, mysterious surface was finally completely mapped. Even better, when the first Surveyor defied popular thinking by landing without being buried in drifts of Moon dust, it showed that it might very well be possible for a *human* to land safely on the Moon.

Manned Space Flight

In early 1961, the Soviet Union was still ahead in the Space Race, as cosmonaut Yuri Gagarin became the first person in space. That was in April; less than a month later, Alan B. Shepard became the first American in space. A year later, the American space program celebrated one of its greatest events. On February 20, 1962, John Glenn became the first American to orbit the Earth. Hailed as a national hero and honored with a ticker-tape parade in New York City, Glenn is still one of the most famous figures of space exploration.

Once manned space flights had begun, programs followed in rapid succession. The Gemini missions carried pairs of astronauts into orbit around the Earth. The Apollo missions were designed to send teams of three astronauts. Unfortunately, the missions were not always successful. The first Apollo launch was marked by a catastrophic fire that killed the three astronauts—Virgil "Gus" Ivan Grissom, Edward Higgins White II, and Roger Bruce Chaffee—in the command module during a launch pad test. As a result of the disaster, the Apollo series was redesigned so that the atmosphere the crew members breathed was not composed of pure oxygen, and the escape hatches were much more accessible and efficient.

Of course, Apollo missions also included great triumphs. Apollo 8 successfully orbited the Moon, paving the way for Apollo 11, when Neil Armstrong and Buzz Aldrin landed on the Moon's surface and walked on the Moon on July 20, 1969. Alas, the manned Moon program ended three years later with Apollo 17 in 1972. Plans had already been made for Apollo missions 18, 19, and 20, which were supposed to be more scientific than their predecessors. But NASA, created to provide civilian control over the space program, cancelled the future missions.

The Moon has never lost its attraction for humans. Japan, China, and India are all developing lunar exploration projects. The United States is restructuring its efforts in human space flight, focusing on developing technologies that will allow a variety of missions beyond low Earth orbit. Missions to the Moon, Mars, and Asteroids will potentially all be possible sometime after 2020.

Twelve Apollo astronauts visited the Moon between 1969 and 1972. In this photo we see Eugene Cernan, the last man to stand on the Moon, during the Apollo 17 mission. The lunar rover's wheel appears at lower right.

Planetary Exploration

The first generation of planetary exploration began with the Pioneer missions. The aircraft were stabilized by their own rotation, which made it hard to take good quality photographs. But despite this complication, Pioneer 10, launched in 1972, brought back very successful studies of Jupiter. Launched a year later, its successor, Pioneer 11, brought us images of Saturn as well as Jupiter. In 1978, heavily clouded Venus became the site of Pioneer missions.

Stabilized on all three perpendicular axes, Mariner spacecraft were capable of much better quality imaging. Mariner 4 flew by Mars in 1965, after a seven-month trip. The problem with this trip was that to the spacecraft cameras flying by, the red planet deceptively looked barren and cratered. In 1971, Mariner 9 became the first spacecraft to orbit Mars. Though it arrived in a dust storm that swept the whole planet, the cameras were able to watch the dust clear and capture very good images of Mars and its moons.

Mars was captured again in 1976 by twin Viking landers. The cameras sent back magnificent images from the planet's surface.

The first panorama transmitted by Viking 1 after its 1976 Mars landing.

Mercury was the target of Mariner 10. Launched in 1974, the spacecraft passed Venus en route, harnessing Venus's gravity to aim it toward Mercury and change its velocity to better match Mercury's. Mariner 10 paid three visits to Mercury in 1974 and 1975 as it orbited the Sun.

Voyager 1 and Voyager 2 were actually Mariner 11 and 12, but the name was changed to make their journeys sound more unique and exciting. (The term "voyager" does seem to invoke the idea of exploring the unknown.) Both Voyagers visited Jupiter and Saturn, and Voyager 2 went beyond them to Uranus and Saturn. The close-up imaging allowed the giant planets and their numerous moons to come into full resolution.

Voyagers 1 and 2 flew by Jupiter in 1979. In 1995, NASA sent the Galileo mission to Jupiter. When it got to the planet, it dropped a probe in the gas giant's atmosphere and continued to send back data to Earth through 2003. That venture ended when it was deliberately crashed into Jupiter's clouds.

Voyager 1 flew by Saturn in 1980, followed by Voyager 2 in 1981. The first spacecraft to orbit Saturn, NASA's Cassini, had its original mission extended and is still orbiting the Saturn system. Hopefully, Cassini will be resilient enough to last a whole season at Saturn, sending us data through 2017 or even beyond that.

Voyager 2 used Saturn's gravity to travel on to Uranus. Then, flying by Uranus in 1986, it used that planet's gravity to turn its path toward Neptune, which it reached in 1989. For fans of these outer planets, unfortunately, for the time being, there are no plans to return.

Pluto was to be the last stop for planetary exploration. The first mission in NASA's new horizon series will arrive at the former planet in 2015. It may be fortuitous that the mission was funded before Pluto's demotion to "dwarf planet." Whether or not Pluto's status might have affected the mission is anyone's guess, but at the time it was part of the mission to reach all the planets, and we may be lucky that Pluto was still the ninth planet.

There are still many missions to other objects in our solar system that are not planets. Halley's Comet was visited by a fleet of missions in 1986. Giotto, sent by the European Space Agency, sent back the best images, with close-up shots of that most famous "dirty snowball."

**The Space Shuttle Discovery blasting off from Cape Canaveral in 2010.
Space shuttle flights ended in 2011.**

NASA's Dawn spacecraft was launched in 2007 with a mission to visit Vesta and Ceres, two of the first discovered and biggest asteroids. Dawn entered Vesta's orbit in 2011, and then a year later flew off on its journey to Ceres. It is expected to reach Ceres in 2015.

MESSENGER, launched by NASA in 2004, stands for MErcury Surface, Space ENvironment, GEochemistry, and Ranging. The first spacecraft to orbit Mercury (whose mission is captured by the somewhat complicated acronym) reached its target in 2011 and accomplished its extended mission in March 2013, successfully imaging all of the small, rocky planet close to the Sun.

Exploring our solar system may have been considered passé for the first half of the twentieth century, but it clearly regained its attraction. Spacecraft with increasingly sophisticated equipment are continually bringing back information on the objects that share our solar system.

PART TWO

Terrestrial Bodies

CHAPTER 4

Mercury: Speeding Around the Sun

In This Chapter

➤ Mercury's high-speed orbit

➤ Observing elusive Mercury

➤ NASA's Missions to Mercury

➤ Mercury in Detail

➤ Future Mercury Exploration

Mercury, messenger of the gods, was depicted with a winged helmet and winged sandals to symbolize his swift movement. Gazing up at the sky, the ancients could see that one of the heavenly "wanderers" moved much faster than its companions with respect to the stars. It was no wonder that the planet seemed to be moving fast. At only 38% of our Earth's distance to the Sun, the small planet named for the Roman god Mercury speeds around the Sun in a mere 88 Earth days.

Mercury orbits the Sun so fast that by the time it has completed a single orbit, the Earth (virtually crawling by comparison) has made only a quarter of its journey around the Sun. Mercury has to continue along its orbit to catch up to the Earth. That period, from our terrestrial perspective, takes 116 days.

Resourceful Copernicus came up with a way to compare the two periods: the one that appears to us from the Earth, and Mercury's real period, as it would appear in space far above the solar system. The period we see from our home planet is called the "synodic period," and the period as seen from above or from the distant stars, is the "sidereal period." As we established in Chapter 1, sidereal means "by the stars." It's a good term to keep in mind—it is commonly used in astronomy.

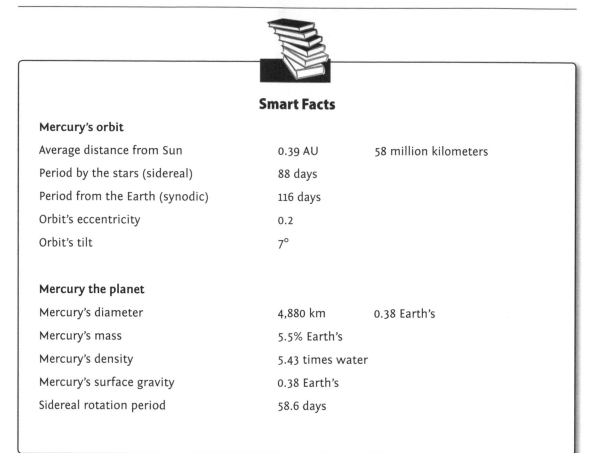

Smart Facts

Mercury's orbit

Average distance from Sun	0.39 AU	58 million kilometers
Period by the stars (sidereal)	88 days	
Period from the Earth (synodic)	116 days	
Orbit's eccentricity	0.2	
Orbit's tilt	7°	

Mercury the planet

Mercury's diameter	4,880 km	0.38 Earth's
Mercury's mass	5.5% Earth's	
Mercury's density	5.43 times water	
Mercury's surface gravity	0.38 Earth's	
Sidereal rotation period	58.6 days	

Copernicus helped us learn more about Mercury, but according to legend, the planet is always so close to the Sun as seen from the Earth that he never got to observe it himself. Centuries later, we have MESSENGER, sending back amazingly detailed images on its extended Mercury mission.

No Goldilocks Planet Here

Mercury's proximity to the Sun—two-and-one-half times as close as the Earth—makes it very hot as well as very fast. During the day, the temperature climbs to a scorching 800 degrees F (roughly 425 degrees C). Then at night, it drops dramatically, to about -280 degrees F (-185 degrees C).

As extreme as those temperatures seem, they are actually moderate compared to what scientists originally thought. The prevailing belief was that one side of Mercury always faced the Sun, locked in by gravity the same way we always see one side of the Moon from the Earth. That was dispelled when the first measurements of radio waves emitted by Mercury revealed that the planet's supposed "dark" side was not quite as cold as it should have been, meaning it must have gotten some light from the Sun. The explanation for this came from radar (which stands for "radio detection and ranging"). Radio waves bounced off Mercury showed that, like our own planet, Mercury rotates with respect to the Sun.

Solar System Scoop

Before Mercury's rotation was discovered, it was thought to be the hottest and coldest place in the solar system. One side was presumed to be relentlessly baking under the Sun's burning rays while the other side would be completely devoid of sunlight and so, always cold. Then radio and radar measurements showed that Mercury rotates so that all parts of its surface are sometimes in sunlight and sometimes dark. We now know that that there are hotter and colder objects in our solar system than Mercury.

Mercury takes 59 days to rotate, a full two-thirds of the 88 days it spends whizzing around the Sun. In other words, it rotates three times for every two revolutions. The Sun's gravity pulling on Mercury's irregular mass keeps it in a gravity lock of sorts, though not as tight as the lock our Earth has on the Moon. If it was possible to stand on Mercury's surface near the equator, we would be drenched in sunlight for the equivalent of 88 Earth days, and surrounded by total darkness for another 88 days.

Mercury Gazing

Mercury literally pales in comparison to Venus (which is ten times brighter), and it can be very tricky to see. At times though, Mercury can outshine even the brightest star in the sky.

Mercury's orbit is inside Venus's, only about 40% of the distance to Earth, compared with 70% or so for Venus. Because of this, Mercury never strays as far from the Sun as Venus. For the purpose of observing the planets, that means that when we watch the two planets setting after the Sun in the western sky, Mercury usually sets first. Reverse that scenario, and when Mercury and Venus rise before the Sun in the east, Venus is usually first. By the time Mercury rises, the Sun is about to light the sky. The best way to spot the elusive Mercury is to find out the best days or weeks to find it.

With the naked eye, or even with binoculars, Mercury looks like a white dot in the sky. Even bright Venus looks the same way. If you want to see phases on the two planets, you need a good (and big) telescope. Mercury, though, is still pretty elusive from Earth. It's much easier to catch the phases of bigger and closer Venus. It's a lot harder to capture the phases of Mercury, and even with a telescope, very few people actually do.

Venus-Mercury Conjunction (April 4th, 8th, 9th, 11th, 12th, 13th, 14th and 15th 2010)
Composite of 8 images taken between 4-15 April 2010 at 19:50 UT. The crescent Moon is from the 15th April image   Pete Lawrence, UK

A series of nightly views at sunset in spring 2010 showed Venus and Mercury in the western sky, with the crescent moon joining the planets on one of the nights in this composite photograph.

Mercury Missions

It is not only amateurs with telescopes on rooftops or in backyards and parks that have difficulty capturing Mercury. Because Mercury is always so close to the Sun in the sky, it is never in complete darkness to us here on Earth. Observing Mercury at twilight when it is low in the sky means watching diagonally though masses of air in our atmosphere. The turbulence in the atmosphere makes the images hazy to even our biggest, most powerful telescopes. It would take cameras on spacecraft to give us detailed images.

Mariner

Mariner 10 headed for Mercury in the 1970s. With the Mariner spacecraft stabilized on all three perpendicular axes, the telescope could take much better and clearer images than the earlier generation of spinning spacecraft. Reaching Mercury for the first time in 1974, the 475-kilogram spacecraft captured close-up images of the planet's surface and even managed to capture images of a sizable part of Mercury as a whole. The problem was that while using Venus's gravity successfully aimed the spacecraft toward its intended target, attempting to match its speed with Mercury's was less successful. One of Mercury's many puzzling characteristics is that it has a lightning-fast revolution but a strangely slow rotation. Mariner went whizzing by Mercury with only a brief time to take photographs.

In orbit around the Sun, Mariner made two more trips around Mercury. Though it successfully photographed more of its surface, when the Mariner mission ended, more than half of Mercury's surface had still eluded its cameras. Most of the small, hot planet was still a mystery.

MESSENGER

It was another thirty years before NASA sent the MErcury Surface, Space ENvironment, GEochemistry, and Ranging to the innermost planet. Weighing about 1,100 kilograms (2,400 pounds), MESSENGER is more than twice as heavy as Mariner. Getting the massive spacecraft to Mercury as it races around the Sun was a tricky venture. To obtain its speed, MESSENGER harnessed the gravity of Earth and Venus. On August 5, 2005, MESSENGER flew by its home planet for a gravity boost and then it flew by Venus twice, on October 24, 2006, and June 5, 2007.

Solar System Scoop

The significance of the word "messenger" goes beyond the planet Mercury's mythological name. We can trace modern astronomy back to Galileo's 1610 book *Starry Messenger*, so the word means a lot for astronomy.

MESSENGER arrived at its destination in 2008, flying by Mercury first on January 14th and then on October 6th, using its gravity to regulate its orbit around the Sun. Its third flyby was on September 29, 2009, gliding over 142 miles (228 km) of Mercury's surface.

The flybys allowed the spacecraft to slow its velocity while conserving fuel. Unlike Mariner, MESSENGER was destined to be orbiting Mercury, taking photographs and sending back scientific data for a long time. MESSENGER entered Mercury's orbit on March 18, 2011. The Mercury Dual Imaging System (MDIS) sent back its first historic photograph on March 29. By April 4th, the data collection process was in full swing.

MESSENGER's original mission was to orbit Mercury every twelve hours for one Earth year. Equipped with an array of sophisticated technologies, the spacecraft was designed to perform the first complete survey of Mercury, including its surface composition, geological history, atmosphere, magnetic field, and plasma environment. Features that appeared in Mariner's hazy images would now be brought into sharp resolution—and that was only part of the venture.

Within six months, there were some exciting new discoveries about the curious innermost planet. The high-resolution cameras revealed clusters of highly unusual depressions and ridges, ranging wildly in size, etched into some of the craters. These unexpected features suggested that Mercury was more volatile than it was thought to be. Its magnetic field, which had always puzzled scientists, was found to have a very distinctive north-south asymmetry. Mercury rocks turned out to be very different in composition from Moon rocks.

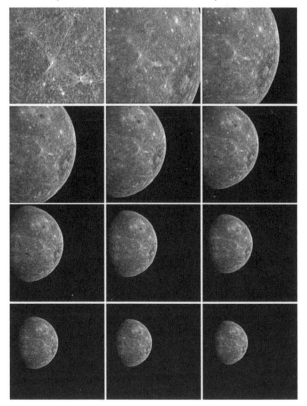

The view as NASA's MESSENGER spacecraft departed from Mercury on January 14-15, 2008, with distances ranging from 21,000 miles to 250,000 miles.

On March 17, 2012, MESSENGER completed its primary mission and NASA could say that the venture was a tremendous success. A year later, they celebrated the successful completion of its *extended* mission. With a final image of Mercury's north polar area, MESSENGER successfully mapped 100% of the planet's small, rocky surface. Further exploration of the unusual craters revealed never-before encountered patterns of "wrinkle ridges" that look like the cracks we see in a pie crust. These are associated with "ghost craters" formed by intensive volcanic activity. Mercury is not only more volatile than scientists thought, but it is also home to rock formations not seen anywhere else in our solar system (at least, not so far).

One of the major discoveries was the presence of water ice at Mercury's frigid north pole. The mission made a tremendous contribution to our knowledge of Mercury (though that admittedly was pretty limited). It also aroused curiosity even more to learn more about the innermost planet. Scientists are now looking forward to the prospect of a second extended MESSENGER mission.

Ready for its Close-Up

So much of Mercury was unknown before spacecraft could be sent there that even discovering that its surface is pockmarked with craters was a remarkable revelation. Low resolution flyby images confirmed their existence and showed some intriguing deposits at the bottom of some of the craters. They also revealed sweeping, fairly flat plains, some of them dotted with cliffs called scarps. Some extend upward as high as a mile and outward for hundreds of miles. But it was not until MESSENGER that we learned just how unique the formations on Mercury really are. MESSENGER's data is still being processed. As more of Mercury's quirks are revealed, they raise even more questions for a second extended mission.

Solar System Scoop

Mercury not only has some of the most distinctive features in the solar system, but its features have a very eclectic collection of names. Some features, like the plains, extend the tradition of using names drawn from mythology. For example, there are plains named after the Norse god Odin, the Persian god Tir, and the Japanese god Suisei. The craters are named for artists, musicians/composers, and authors. Mozart and Bach are two of the first named craters. MESSENGER's journey inspired the naming of nine additional craters. In this group, the Roman poet Catullus is joined by figures from the 19th and 20th centuries that include a film composer (Komeda, for Polish film music composer and jazz pianist Krzysztof Komeda), the father of modern blues (Waters, for McKinley "Muddy Waters" Morganfield), and the visionary filmmaker and animator who always thought that one day space travel would be a reality (Disney, for Walt Disney). The scarps are named for famous ships that explored the world. This group includes Columbus's *Santa Maria*, Captain Cook's ship *Endeavor*, and Darwin's ship *Beagle*.

The Moon has often been used as a reference point for trying to understand Mercury. Mariner disproved the idea that one side of Mercury was always dark. How different would Mercury prove to be in other ways? We've known about the Moon's craters since Galileo. Today we have photographs of the Moon's craggy surfaces taken from numerous angles, from telescopes on the ground, telescopes in space, and by the astronauts' handheld cameras. Even before MESSENGER, it was obvious that Mercury's craters were subtly different from the craters dotting the lunar landscape. Mercury's gravity is higher than the Moon's, and the rims of its craters are thinner and flatter.

MESSENGER shows us in detail how cratered Mercury really is. Impact-wise, Mercury evolved in a dangerous neighborhood. On both Mercury and the Moon, the craters were probably formed by meteorites. To some extent, the Earth's gravity shielded the Moon from the impact of hurtling meteorites. Mercury was clobbered by meteorites in full force! Some of the impacts were massive. One such impact undoubtedly formed the immense Caloris Basin, and its shock waves extended around the planet and formed what is described as the "weird terrain" on the side of Mercury opposite Caloris.

When one crater appears on top of another, it's easy for us to date them in relation to one another, since it's obvious that the lower one must have been formed first. But that does not allow us to tell their absolute ages. We can date the rocks and dust that the astronauts brought back from the Moon, but we still have no direct measurements of the ages of Mercury's craters.

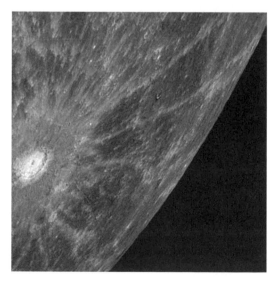

Rays from a relatively recent crater in Mercury's southern hemisphere. This crater and the rays were first detected over 40 years ago with radar. The crater is about 50 miles in diameter.

On Mercury, as on the Moon, some craters give off bright rays. This shiny stuff is newer material from under the planet's surface that was ejected during the impact. Over millions of years, it gets darker. The systems that give off rays must have been formed in the last hundred million years or so. That's only a mere 2% of the age of the solar system—recent by astronomical standards.

The craters might have been formed by an ongoing invasion of tiny meteorites, or "micrometeorites." Alternately, Mercury's surface might have been softer, which would have allowed the rims of the craters to settle. Or, in another scenario, Mercury's gravity might have compressed the rims to some degree, compared to our gravity-less Moon.

Mercury's scarps are another point of curiosity. How did those lines of cliffs come to be? It is probable that Mercury shrunk, perhaps by a mile or somewhat longer (one or two

kilometers), while it was in a molten state, and then cooled. Mercury's scarps are much more widespread than those on our home planet. Scarps are particularly prevalent around Mercury's south pole.

Mercury's west limb, with north at the right, shows many craters flooded with lava and large regions of smooth plains, caused by vulcanism. There are also lines of cliffs known as scarps, which can extend hundreds of miles across Mercury's surface.

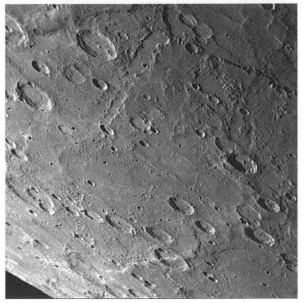

This is the impact basin discovered on MESSENGER's second flyby of Mercury which is named after the artist Rembrandt. Two scarps, lines of cliffs, cut across craters that are older.

Inside the Innermost Planet

It was quite an unexpected discovery when Mariner 10 detected a magnetic field on the innermost planet. Magnetic fields are thought to be generated by the circulation of molten materials deep in the planets' interiors. The popular opinion was that Mercury was so small that it would have cooled completely inside the planet's core.

Mercury's magnetic field is detectable but it is still very weak—only about 1% as strong as our Earth's. It might be left from an early stage of Mercury's evolution, when the interior was still molten, and is gradually dissipating. For now, it is probably still being generated in a shell of molten liquid within a solid core. The north-south asymmetry of Mercury's magnetic field is one of MESSENGER's many discoveries that are still being analyzed.

Though much smaller than Earth, Mercury has similar density, meaning it must have a core of heavy elements composing about 60% of the small planet's mass. That's roughly twice as much as our home planet's iron-rich core makes up of its mass. Mercury's core accounts for about 75% of its diameter. MESSENGER's discoveries about Mercury's composition have raised even more intriguing questions about the planet. If NASA approves a second extended mission, it should provide us with new insight into the planet's unusual composition and structure.

What's in an Atmosphere?

Atmospheres are composed of tiny gas particles zooming around. Our own atmosphere consists of about 80% nitrogen molecules and 20% oxygen molecules, with a lot of other elements in the mix. The faster the particles speed around, the hotter the atmosphere feels. We routinely talk about the temperature outside or inside our home, but technically, "temperature" applied to a gas means the average speed of its atoms or molecules.

Our Earth's gravity locks in the molecules at the speeds they travel. Mercury though, has a much higher temperature since it's so near to the Sun, and it has less gravity at its surface. (Surface gravity depends on the amount of mass enclosed and how far the surface is from its center. Being much less massive than Earth but not that much smaller to counterbalance the lower mass, Mercury's surface gravity is roughly 38% of Earth's.) Mercury is bathed in solar radiation. With so much sunlight per square inch (per square meter), it was always thought to be much too hot to retain an atmosphere.

That long-held belief was dispelled by Earth-based observations detecting sodium and potassium gas around Mercury. It turned out Mercury does have an atmosphere, though it is very, very thin. Mercury's atmosphere contains about 150,000 atoms of sodium per cubic

centimeter, many millions of times lower than the
density of gas in our home planet's atmosphere.
Mercury's atmosphere has about 4,500 atoms
per cubic centimeter of helium, combined with
smaller amounts of oxygen, hydrogen, and
potassium. One theory is that some of the gas
may emanate from under the planet's surface
and remain there for awhile before most of the
gas molecules escape. This theory comes from
the association of larger amounts of gas with the
Caloris Basin. It is also possible that some of the
gas comes from the Sun, or was ejected from
those meteorites battering Mercury. Individual gas atoms don't linger very long, so there has
to be a continual supply.

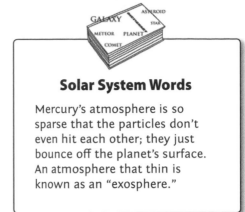

Solar System Words

Mercury's atmosphere is so
sparse that the particles don't
even hit each other; they just
bounce off the planet's surface.
An atmosphere that thin is
known as an "exosphere."

BepiColombo

As MESSENGER orbits Mercury, the European Space Agency, in a joint venture with the
Japan Aerospace Exploration Agency, has been planning its first mission to Mercury. Named
BepiColombo, the mission carries two orbiters. ESA's Mercury Planetary Orbiter will be
exploring Mercury's surface and composition. The Mercury Magnetosphere Orbiter, under

Smart Facts

BepiColombo is named for the 20th century Italian scientist Giuseppe Colombo (1920–1984),
who conducted detailed studies of Mercury and advanced our knowledge of how to travel to
Mercury and other planets. Colombo's nickname was Bepi.

the auspices of the Japanese agency, will study the planet's magnetic field. BepiColombo is
scheduled to launch in 2015 from ESA's spaceport in Kourou, French Guiana. The orbiters
are expected to reach Mercury in 2022.

CHAPTER 5

Venus: The Runaway Greenhouse

In This Chapter

➤ Observing Venus from Earth

➤ The surface from hell

➤ The runaway greenhouse effect: Not the twin we want to be like

➤ Rotating backward

➤ Venus from the ground and the sky

➤ Rare pairs of transits

Staring up at the sky, it is easy to see why Venus would be associated with the goddess of love and beauty. Dazzling brilliant, Venus is both the "morning star" for early risers looking eastward and the "evening star" as the Sun sets in the west, depending on where it is in its orbit around the Sun. However, Venus's beauty is very deceptive. What we don't see from the Earth is that our nearest neighbor—which has often been called Earth's twin—is a raging inferno hot enough to melt lead with a lethal atmosphere.

Like Mercury, Venus's orbit is inside Earth's, so it can cross in front of the Sun. This is called the transit of Venus, a rare and spectacular event, which we will get to later. Though Venus would undoubtedly be stunning against a midnight sky, it is too close to the Sun in the sky to be out that late. At the same time, its orbit is closer to Earth's and bigger than Mercury's, which allows it to move farther out. At its maximum angular distance from the Sun, called "elongation," Venus can be, at maximum, between 45 and 47 degrees from the Sun in the sky. In other words, it can be halfway up in the western sky after sunset, and seen for hours before it finally sets. The maximum elongation is variable because the orbits of the planets are ellipses (remember Kepler's laws from Chapter 2), if they were perfect, concentric circles, the elongation would be the same always.

Gazing through his telescope, Galileo saw Venus go through a full set of phases, which he recorded. To see that from Earth meant that Venus had to be moving around the Sun and not around us. As a key part of Galileo's evidence for a heliocentric universe, the phases of Venus have a prominent place in the history of modern astronomy. MESSENGER saw only one of Venus's phases—a crescent—as it flew by (Figure 1). Watching from Earth, we can see the gibbous (more than half) and almost full phases as well.

Smart Facts

Venus's orbit

Average distance from Sun	0.72 AU	108 million kilometers
Period by the stars (sidereal)	225 days	
Period from the Earth (synodic)	584 days	
Orbit's eccentricity	0.01	
Orbit's tilt	3.4°	

Venus the planet

Venus's diameter	12,104 km	0.95 Earth's
Venus's mass	82% Earth's	
Venus's density	5.24 times water	
Venus's surface gravity	0.89 Earth's	
Sidereal rotation period	243 days retrograde (backward)	

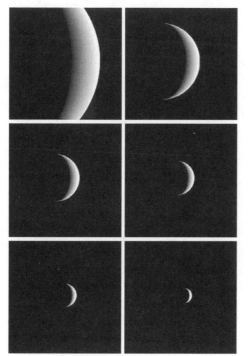

A crescent phase of Venus, from NASA's MESSENGER as it departed from Venue en route to Mercury.

Solar System Scoop

Only the Sun and the Moon can outshine Venus as the brightest celestial objects. Morning star or evening star, Venus's light is strong enough to cast (faint) shadows. But with its orbit within Earth's, it can never stray more than 45 to 47 degrees from the Sun and is never up in the sky very late at night. As Galileo found, even a small telescope can reveal the planet's phases, but details of its surface are hidden.

Hot, Hot, Hot

Venus's thick cloud cover obscures its details from view. Although optical instruments cannot get through the dense cloud cover, radio waves easily can. When radio telescopes were first used to measure emissions from Venus, that part of the spectrum was amazingly bright. The brightness revealed that the temperature—on the night side as well as the day side—was 900 degrees F (750 kelvins, or 475 degrees C). That even surpasses Mercury at its hottest!

We know Venus is closer to the Sun than we are, which raises the question of how hot we might expect it to be. If Venus had no atmosphere, we could calculate how much energy the Sun gets per second. That result is simply the percentage of the angle of the sky that is occupied by Venus's disk. We can also calculate how much energy Venus would emit, if it balanced the energy taken in with the energy it gives out. That amount follows a formula commonly used in physics. The energy increases by the fourth power of the temperature, using temperatures based in absolute terms. The temperature scale for this is the kelvin scale, which begins at absolute zero = -273 degrees and escalates with the same size degrees as the Celsius scale.

Solar System Words

The unit of heat is officially written as "kelvin" with a lower case "k," for example 273 kelvin. But since it is named after a person, Lord Kelvin, its symbol is the capital letter "K" and when scientists refer to the Kelvin scale, they use the capitalized word; as in the Kelvin scale. Also, unlike with the Celsisus and Fahrenheit scales, the word "degrees" is never used when using the Kelvin scale.

Note the fourth-power dependence in the formula. That means that even a slight change in temperature can have a huge effect on the amount of energy emitted. For example, though we are not even close to a change by a factor of 2, a change in temperature by a factor of 2 produces 24 = 2×2×2×2 = 16 times change in energy. A change by 10% is a factor of 1.1, and 1.14 = 1.212 = 1.45 in energy, so that even a 10% alteration in temperature results in a 45% change in energy.

If we balance energy in and energy out, the resulting calculation is a mere 215 degrees F (375 K = 100 degrees C). That's barely one-quarter of Venus's scorching temperature! Even though it is much closer to the Sun than we are, Venus is a lot hotter than we would expect it to be with a conventional formula.

The reason for Venus's searing heat is known as the "greenhouse effect" (though we have since discovered that our own greenhouses are heated by a different mechanism). What it means is that Venus's atmosphere is transparent or translucent in the visible part of the spectrum where most of the Sun's energy is concentrated. That atmosphere heats the surface to a temperature that emits rays primarily in the infrared where Venus's atmosphere is mainly opaque. So the surface of Venus is forced to keep heating up (and heating up, and heating up) until it gets hot enough that even the tiny percentage of energy need to balance the incoming energy can escape.

For a long time, the popular wisdom about our greenhouses here on Earth was that the glass caused a similar type of greenhouse effect. It was finally discovered that even if you build a greenhouse out of solid salt, which is transparent in the infrared, its interior will heat up anyway. Most of the heating inside a greenhouse is due to the fact that the glass or other materials keeps breezes from entering and distributing any hotter air.

On Venus, the greenhouse effect is caused by its very thick atmosphere of carbon dioxide, which gives it an atmosphere about 100 times thicker than ours. CO_2 accounts for 96% of the mass of Venus's atmosphere. One term used to describe this phenomenon is a "runaway greenhouse effect." It ensures that our twin planet is completely uninhabitable for human beings.

Solar System Scoop

The term "runaway greenhouse" became symbolic of a worst-case scenario for Earth as evidence of human-caused climate change built up. Venus's runaway greenhouse effect is a key factor in scientists' fears about the effect of our adding more and more carbon dioxide to our planet's atmosphere. A small greenhouse effect on Earth actually keeps our home planet habitable, but the increase predicted by our current rate of CO_2 emissions from burning fossil fuels and other sources could be catastrophic. While we may be a long way from Venus's runaway greenhouse effect, we don't want to see our planet move even a little bit in the direction of becoming like our inhabitable twin. We will discuss this problem in more detail in the chapter on the Earth.

How did Venus come to be enshrouded in carbon dioxide? It is possible that Venus might have once had a cooler climate, perhaps even Venusian oceans. (Some scientists even wonder if life forms could have evolved under those more idyllic conditions and possibly some even survived, however extreme and inhospitable the planet might be.) Then at some point, volcanoes might have spewed carbon dioxide into the atmosphere. As carbon dioxide built up, the temperature would have soared and the oceans evaporated, leaving the planet with no way to absorb carbon dioxide. The water vapor from this effect could have broken down into its elements and the hydrogen atoms, with their relative lightness, could have speeded up and away from Venus's gravity.

On our Earth, a lot of the carbon dioxide was infused into rocks. Venus might have become very hot at a very young stage of its life. With no oceans to help the rocks absorb carbon dioxide (that is, no carbon cycle like the one we have here on Earth) the runaway greenhouse effect continued to escalate. Fortunately for us, there are numerous differences in the way Venus and the Earth evolved despite similar size, gravity, rocky terrain, and proximity. Comparing the two planets, formally known as comparative planetology, provides us with valuable insight for conserving our own atmosphere and seeing to it that our future does not hold a runaway greenhouse effect.

Moving Backward

In addition to revealing that Venus is bathed in heat, radio telescopes showed that Venus is unique among solar planets in its rotation. Of all the planets, Venus has the least eccentric orbit—and the most unusual rotation. Venus has the slowest spin on its axis of all the planets, but not only that, it rotates backward. While its seven solar companions rotate in a counterclockwise direction, Venus alone has a clockwise spin. The term for this backward rotation is retrograde, analogous to the apparent backward or retrograde movement of the planets across the sky.

Smart Facts

Rotation and revolution are both aspects of the phenomenon of "angular momentum." If you wrap one hand around a spinning object, your upraised thumb points in the same direction as its spin or revolution. Angular momentum is typically conserved, meaning that neither the degree of spin of rotation nor the direction of the spinning object will change by itself. We call this the law of conservation of angular momentum, with conservation implying that there is no change.

In view of the fact that angular momentum is ordinarily conserved, theoretically, Venus should be rotating on its axis in the same direction as all the other planets as it revolves around the Sun. That is, unless some external force altered its spin. The dominant theory is that early on its evolution, Venus was hit by a huge object. As we discussed with Mercury, the early solar system was a very chaotic place. Numerous objects collided and numerous objects were deflected by the gravity of other objects. A big enough object hitting Venus could have thrown its rotation into reverse. There are also other ideas involving tidal effects in the planet's very dense atmosphere. Rising tides caused by slight changes in Venus's distance from the Sun could have been pulled on by the other planets' gravitational forces.

The Dense Clouds Uncovered

Seen from Earth through ordinary light, Venus is a gleaming object going through phases like the Moon. Most of the sunlight reaching Venus is reflected. (The percentage of light reflected by a body is called its "albedo.") Then, there is that dense cloud cover that keeps Venus's surface from view. The thick clouds are composed primarily of droplets of sulfuric acid. There are water vapor droplets mixed in, but like the rest of the planet, Venusian rain is highly toxic. The clouds let in only a scant 2% of the incoming light. The only light that gets through Venus's thick atmosphere is reddish and faint—a startling contrast to the dazzle we see from the Earth.

Ultraviolet light makes Venus's clouds visible; even the longer ultraviolet rays close to the visible violet rays of the spectrum are sufficient to bring them out. Thanks to the many American and Russian spacecraft that have flown by and orbited Venus, we have numerous photographs of Venusian cloud circulation. Venus's location as a point where spacecraft could get a boost on route to other destinations actually has an added advantage—the space telescopes can capture image of Venus's clouds and other features as they fly by. As it orbits the Earth, the Hubble Telescope also snaps photographs of Venus's clouds.

Like its rotation, Venus's clouds also move backward. Only unlike the planet's sluggish rotation, the clouds move rapidly. They take only four days to go around, roughly sixty times faster than the rotation of Venus's surface. If we could see them from Earth, we would probably be fascinated by its cloud patterns that change over hours.

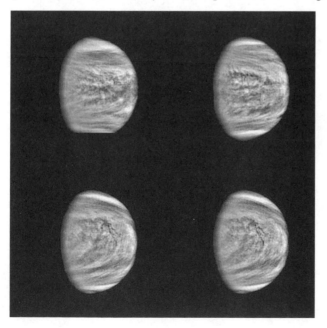

A series of photographs of Venus's clouds, taken in the ultraviolet. The images were taken at intervals of a day in the top row and two in the bottom row. The bright polar hoods are common on Venus, but cloud patterns often change at lower latitudes. The photographs were taken from NASA's Galileo mission, as Galileo departed en route to Jupiter.

Smart Facts

Flying to, Flying by Venus

Venera 7 (Soviet Union), 1970, landed, survived for 23 minutes

Venera 8 (Soviet Union), 1972, landed, survived for 50 minutes

Pioneer Venus (U.S., NASA)

Magellan (U.S., NASA), 1990-93, orbiter

Venus Express (European Space Agency), arrived 2006

MESSENGER (U.S., NASA), flybys in 2008 and 2009 en route to Mercury

Akatsuki (Daybreak) (Japan), launched May 2010 to conduct a thorough exploration of Venus and its atmosphere

BepiColombo (European Space Agency and Japan Aerospace Exploration Agency), scheduled for launch 2015, two flybys of Venus en route to Mercury

Venus Seen from the Ground

While the United States was focused on the Moon, the Soviet Union in the 1960s and 1970s made Venus a top priority. As part of their major space project, they undertook the challenging endeavor of actually landing a spacecraft on Venus's very inhospitable surface and having it last long enough to send back usable data despite the searing temperature and toxic atmosphere.

There were finally several brief but successful landings. In 1970, Venera 7 survived on Venus's surface for 23 minutes before it succumbed. Two years later, in 1972, Venera 8 lasted for 50 minutes. Using the dim orange light of the planet's surface, the spacecraft sent back images of the Venusian landscape showing flat rocks that look very similar to common basaltic Earth rocks. Measurements of the soil's composition and density also revealed notable similarities to the soil here on Earth, as well as on Mars and the Moon. The similarities ended there. The data confirmed the high temperature and density that distinguish the surface of Venus. Before the Soviet space mission, these measurements had only been made remotely from Earth.

What would winds be like on a scorching hot planet with a thick atmosphere? On the descent though Venus's atmosphere, the spacecraft transmitted data on the winds at different levels. The winds are fairly slow and close to the surface. But slow or not, these winds are powerful, sending dust and pebbles flying across the surface. The spacecraft also detected three levels of clouds, separated by relatively clear areas.

Venus Seen from Above

Radar waves have proved to be invaluable for revealing the secrets of Venus. With radio instruments that penetrate the thick clouds, spacecraft orbiting Venus send back data mapping the Venusian terrain. NASA's Magellan spacecraft did a great job of that in the 1990s, sending back high-resolution images that mapped virtually all the surface of Venus.

There have been images taken by the Arecibo radio telescope and other giant radars here on Earth (using techniques we discussed in the Mercury chapter), but they could only capture small portions of Venus's surface. The orbiting spacecraft captured the global panorama. On earlier missions, NASA's Pioneer Venus spacecraft and the Russian Venera 15 and 16 used radar to map the whole surface.

Images sent by Magellan revealed that most of Venus's surface is covered by rolling plain. The surface is so flat that there is no area much more than a mile from the average level. While on Earth we have deep, wide-sweeping ocean basins covering most of our planet's surface, on Venus only 16% of the surface drops below average. Of course, there are no oceans on Venus to cover the deep depressions, one of the many ways in which our sister planet differs from us. Figure 3 shows Venus's topography derived from radar imaging.

A radar image of Venus, with the northern continent Terra Ishtar and the larger equatorial continent Terra Aphrodite displaying as bright radar reflections.

A perspective view of what flying over Venus surface might be like compiled from radar data from NASA's Jet Propulsion Laboratory in Pasadena with the altitudes exaggerated by a factor of about twenty to magnify the height differences of the features. The volcano Gula Mons appears on the horizon on the right about 1.8 miles high.

Venus has two elevated regions that correspond to our continents. In the north, the continent Terra Ishtar is roughly the same size as the United States (for perspective, remember that Venus about the same size as Earth). This continent is home to the highest point on Venus: Maxwell Montes. At 11 kilometers in altitude, this peak surpasses Mount Everest in height. Maxwell had appeared as a bright spot on terrestrial radar imaging, but it was not until Venus was fully mapped that it was revealed as a mountain pointing up from a high plateau.

Smart Facts

Maxwell Montes was named for James Clerk Maxwell, the renowned physicist who unified electricity and magnetism. Maxwell, the scientist, stands with Isaac Newton and Albert Einstein as the three greatest physicists of all time. On Venus, Maxwell Montes is the only feature named after a male. All other Venusian features have female names.

The second continent on Venus, Aphrodite Terra, is twice as big as Terra Ishtar and located near the equator. Aphrodite is Venus's Greek counterpart as the goddess of love and beauty. Ishtar is the Babylonian goddess of love and fertility.

The global maps of Venus show no evidence of the plate tectonics, that is, moving continental regions known as plates that we have here on Earth. So unlike our earthly continents, which can shift over time (some time in the distant future, California may be separated from the rest of the continental U.S.), Ishtar and Aphrodite will stay where they are now.

Solar System Scoop

In May 2010, the Japan Aerospace Exploration Agency (JAXA) launched Akatsuki with the goal of leading a new era in the exploration of Venus. The spacecraft had a smooth start, but failed to enter Venus's orbit in 2011. Several adjustments were made, and the spacecraft is expected to enter Venus's orbit in 2015 or 2016. Akatsuki will be using infrared rays to study the features of the Venusian surface and atmosphere that still remain a mystery.

Transits of Venus

The transit of Venus is one of the rarest celestial events. Those of you who were lucky enough to see one (or both, if you were very fortunate) of the last pair of transits of Venus in 2004 and 2012, were probably thrilled by the spectacular event. If you missed them, the good news is that there are many excellent images available on the Internet. The bad news is that the next pair of transits will not be happening until December 11, 2117, and December 8, 2125.

Johannes Kepler was the first person to predict the transits of Venus, in his Rudolphine Tables of 1627. In 1639, a young English astronomer named Jeremiah Horrocks elaborated on Kepler's predictions and became one of two people to actually see the elusive event—the other one being someone that Horrocks corresponded with. Silhouetted against the Sun, Venus turned out to be a lot smaller than Horrocks expected.

Astronomers in the Age of Enlightenment relied on Kepler's laws of planetary motion for the relative sizes of the planets' orbits. They had no way of measuring the absolute sizes until the distance from the Earth to another planet was measured directly so a full scale of the solar system could be devised.

Transits of Venus were thought to hold the key to a perennial problem puzzling astronomers: the size of the solar system and the distances of the planets. This theory carried its own problem, namely that transits of Venus are rare events. They happen in pairs separated by eight years, but then it takes more than one hundred years for the next pair. The transit witnessed by Horrocks was the second one of a 1631/1639 pair. The next transits of Venus would be in 1761 and 1769.

During this time, Edmond Halley, best known for the famous comet, proposed a strategy for calculating the size of the solar system. His proposed method was to time how long it takes Venus to cross the face of the Sun from various latitudes, therefore enabling the use of triangulation, similar to what is done in surveying, to figure out how far away Venus is.

Halley's method would work, but with one tricky condition: the transits would have to be timed to about one second. As it turned out, when the 1761 and 1769 transits were observed, Venus did not separate cleanly from the edge of the Sun when it moved inside the Sun. There was a black region connecting the black outside of the Sun's disk with Venus's black circular disk. As Venus moved farther in, that black area stretched and eventually "popped." By the time Venus was clearly inside the solar disk, a minute had passed. The same phenomenon happened again when Venus was leaving the solar disk. The result was that the measurements were accurate to about one minute instead of the single second that Halley's method required. Halley's method demanded too much precision to provide an accurate scale of the solar system. The black region joining Venus and the edge of the Sun came to be called a "black drop," and the problem got the name the "black-drop effect." This effect is now known to be due to the combination of how telescopes form images and the dramatic darkening of the Sun towards its edge, called "limb darkening".

For the next transits of Venus, astronomers had to wait until 1872 and 1884. Photography had been invented by then, which added to the attraction of seeing the transit. Numerous expeditions set out across the world. Once again, the astronomers were thwarted by the black-drop effect. Ultimately, triangulating with an asteroid that came relatively close to Earth proved to be a more practical and successful method for calculating the size and scale of the solar system.

The last pair of transits of Venus took place on June 8, 2004 and June 5-6, 2012. The beginning of the 2012 transit was visible in North America, the Caribbean, and northwestern South America on June 5th until sunset. People in the Middle East, South Asia, eastern Africa, and most parts of Europe had their turn to see the transit from the time the sun rose on June 6th. It was not visible from western Africa or most of South America. In the U.S., sky watchers in Alaska and Hawaii could see the entire transit including both black drops. Waikiki Beach was a very popular viewing spot, and the University of Hawaii set up eight telescopes and two giant screens showing webcasts of the event. Outside of the U.S., the entire transit was visible in Siberia, eastern Australia, and eastern China.

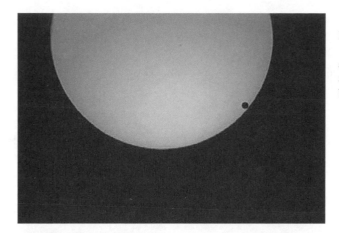

Photo of Transit of Venus of June 8, 2004, with Venus approaching the edge of the Sun taken from Luxor, Egypt, by the author.

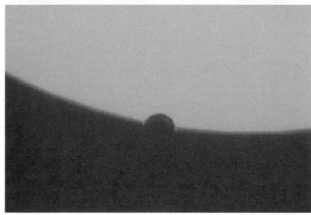

Photo of Transit of Venus of June 8, 2004, with Venus egressing the edge of the Sun taken from Luxor, Egypt, by the author.

Solar System Scoop

We now know that Venus has no moon, but in the 17th and 18th centuries, even noted astronomers insisted they saw one. The first alleged sighting was in 1645 and such sightings continued for the next 123 years. When there were no signs of a moon during the transits of Venus in 1761 and 1769, most astronomers discarded the idea. The public took longer to let it go. As late as 1877, Jules Verne wrote about traveling to Venus with the goal of revealing that "beyond a doubt the planet has no moon or satellite, such as…astronomers have imagined to exist."

 Earth: The Original Goldilocks Planet

In This Chapter

➤ Oceans deep, mountains high

➤ Monitoring our planet from space

➤ Taming carbon emissions

➤ Polluting the night sky with light

➤ The tides and the Moon

The discovery of extrasolar planets has generated a search for a "Goldilocks planet" that is not too hot, too cold, or otherwise too extreme, but is "just right" for hosting life. So far, only one planet seems to fit that description: our own planet Earth. Even with some extreme climate zones, our terrestrial home is temperate enough to live on, and the 20% oxygen in our atmosphere makes it easy for us to breathe.

Oceans cover most of the Earth's surface. Though that adds to our planet's habitability, it complicates the ability to map its topography. We actually have a more accurate picture of the landscapes of various other planets and their moons than we have of the Earth.

Our terrestrial perspective is evident in the way we judge the altitude of our mountains. We think of Mount Everest on the Nepal-China border of the Himalayas as the highest mountain peak on our planet. And the towering peak—the mountain climber's dream—does reach the greatest height above sea level. The term "sea level" is the clue. If you change the vantage point and measure from the bottom of the ocean floor, the tallest peak is Mauna Kea on the island of Hawaii, rising 30,000 feet (10,000 meters) from its watery base. Only 13,796 feet (4,205 meters) of Mauna Kea is above sea level (less than half of Mount Everest); most of it is submerged.

Solar System Scoop

Mount Everest was remeasured in 2005, and the world's tallest peak turned out to be even taller. The new measurement added 14 feet (4.5 meters) to the previous figure of 29,017 feet (8,844.43 meters) or 5.5 miles (8.8 kilometers).

Mauna Kea ("White Mountain") may be dwarfed by Mount Everest when measured conventionally, but it has a very important place in astronomy as an exceptional site for astronomical observation. Also located on Hawaii, Mauna Loa, slightly lower than Mauna Kea at 13,679 feet (4,169 meters), has been providing us with data on carbon dioxide for the last half century that makes a compelling case for changing some of our human activities. We will get to that later in this chapter.

Smart Facts

Earth Facts

Earth's orbit

Average distance from Sun	1 AU	93 million miles = 150 million kilometers
Period by the stars	3651/4 days	
Period from the Earth	--	
Orbit's eccentricity	0.02	
Orbit's tilt	–	

Earth as a planet

Earth's diameter	12,756 km
Earth's mass	5.97 trillion trillion kilograms (6 with 24 zeroes)
Earth's density	5.5 times water
Earth's surface gravity	32 feet/second/second = 9.8 meters/second/second
Sidereal rotation period	23 hours 56 minutes 4 seconds

Putting Earth in Perspective

Seeing things in perspective requires some distance. Before the space program, it was impossible to see the Earth as a whole. The photographs taken by the Apollo 8 astronauts as they flew from the Earth to the Moon in 1968 gave tremendous momentum to the ecology movement.

The cost of the space program has been controversial for years, but it has more than repaid itself with technologies for communication and for monitoring our planet's resources. Along with NASA, the U.S. National Atmospheric and Oceanographic Administration (NOAA) has a series of spacecraft tracking the Earth from orbit. One of its major spacecraft series is the Geostationary Satellite Server (GOES). In December 2009, GOES-14, the most advanced in the series, became the main GOES satellite, measuring x-rays, protons, and electrons in Earth's orbit, as well as its magnetic field. The GOES series is powered by solar cells. Scientists recognized that because its solar cells face the Sun, the GOES satellite could be equipped with another telescope for monitoring x-rays emitted by the Sun. From its vantage point above the Earth's atmosphere (which absorbs x-rays and so, protects us from them), it transmits x-ray data about solar storms and warns us about any potential dangers caused by those storms to our planet or our satellites.

Most of the GOES satellite data gets sent to the Weather Prediction Center of the National Weather Service. You can find its Space Weather Now page online at http://www.swpc.noaa.gov/SWN/.

1992's Hurricane Andrew as a spiraling storm in the Gulf Coast from the National Oceanic and Atmospheric Administration's GOES-7 spacecraft.

Keeping Track of the Oceans

Popular images of the Earth taken from space clearly show that we live on an ocean planet. Since we live surrounded by oceans, it is in our best interests to learn all about them, from the depths of the ocean floor to the highest waves. Wave heights have been charted by a series of oceanographic satellite maps. In recent years, the term *El Niño* entered the popular lexicon as we became aware of the powerful effects of this phenomenon—a warming in the equatorial Pacific—on our weather. (El Niño, meaning a male child, refers to the Christ Child because it often starts around Christmastime.) Flooding and drought are both weather extremes wrought by El Niño. In the southern U.S., for example, flooding resulted from heavy rainfall caused by El Niño, while Australia was hit by drought, which produced raging brush fires. The trade winds are weakened, and a prolonged El Niño can be devastating to the fishing industry.

All these effects of El Niño are monitored by NOAA, which keeps track of sea temperatures, currents, and winds from buoys in the sea as well as from space. We can expect El Niño roughly every three or four years, though it does not have a regular, predictable pattern.

NASA's Jason-2 satellite keeps track of sea level height and ocean temperatures. In 2009, a series of images revealed warm water moving eastward toward South America. The warm stream brought rainfall and floods to Peru, but left Australia and Indonesia with drought. With a constantly changing pattern of atmospheric circulation, the consequences of El Niño extend way beyond its original ocean source.

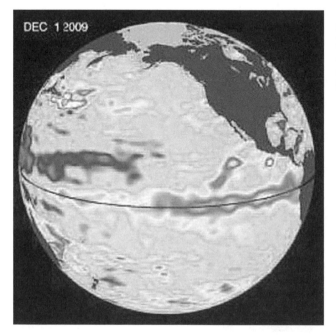

Sea level height, imaged from the Jason-2 oceanographic satellite. Along the equator, there is a strong wave of warm water that is approaching Peru in South America.

El Niño has a counterpart, known as La Niña (female child), a cooling in the equatorial Pacific. La Niña often follows El Niño, though that is not always the case. Jason-2 maps La Niña and El Niño.

The Southern Oscillation Index captures the full range of variation in the southern Pacific Ocean: a difference in surface pressure between French Polynesia and Darwin, Australia. The index is linked with the strength of the trade winds, which flow in part from high pressure regions to low pressure regions. El Niño and La Niña are both manifestations of the El Niño Southern Oscillation, as a warm and cool ENSO, respectively.

Terra Firma and its Interior

Earthquake waves on our planet's surface provide geologists with important clues to the Earth's interior. The interior is divided into three regions: a central core, a surrounding mantle, and a surface crust. The upper mantle and crust are both solid, while the lower mantle is partially melted. This composition is not unique to our Earth; the other terrestrial planets and the moons circling the outer planets are similar in structure.

The Earth has a molten, metallic core, whose rotation creates the magnetic field. Our north magnetic field is presently located in Hudson Bay, Canada, and is traveling northward at roughly 10 miles (15 kilometers) per hour.

The crust is made up of continental plates that are expansive in breadth but thin in depth. Plate tectonics refers to the study of the plates and their movements. In the primeval days of our planet some 270 million years ago, these plates formed one unified continent known as Pangaea. Ever since they separated, they have been moving apart and continue their drift over the Earth. (They move at about the same rate as our fingernails grow.) Earthquakes often happen at plate boundaries, such as the San Andreas Fault in California, triggered by the actions of the plates pushing against each other, or under and over each other.

Solar System Words

Tectonics is the branch of geology that examines the Earth's crust and the type and distributions of its rocks and mineral. The term "tectonics" comes from the Greek word for builder.

The notorious San Andreas Fault is the plate boundary between the California (Pacific) plate and the North American plate, encompassing everything to its east. Someday California will be separated from the rest of the continent. Other objects in our solar system, such as Jupiter's moon, Europa, experience similar tectonic drifting.

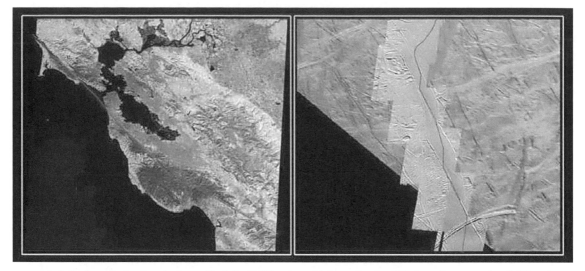

At left, there is the San Andreas Fault in California, which marks the difference between the California (Pacific) plate and the North American plate. At right, there is an image of a similar type of fault, known as a slip-strike fault, on Jupiter's moon Europa. A line is drawn in to mark the fault's position. Both images are shown at the same scale and resolution.

From Global Warming to Global Climate Change

Climate skeptics are often quick to seize on unusually cool or cold weather as evidence that global warming is an elaborate "hoax." (They also confuse "climate" with "weather"; weather refers to atmospheric conditions over a short duration, while climate refers to the average of day-to-day weather over an extended time period.) Global climate change is a more accurate term for what is happening to our planet, and something on which virtually all scientists agree is indeed happening.

As we discussed in the chapter on Venus, we do not want to become anything like our uninhabitable twin. Venus's runaway greenhouse effect has left scientists concerned for the fate of our own planet. The concentration of carbon dioxide on Venus, about 100 times Earth's, combined with Venus's closer distance to the Sun (70% of ours) has generated a tremendous increase in Venus's surface temperature. Being the "right" distance from the Sun is part of Earth's status as a Goldilocks planet, but that does not protect us from an increase of carbon dioxide in our atmosphere as a consequence of human activity.

On Earth, as on Venus, the greenhouse effect, in which sunlight can reach the surface but the infrared emitted by the warmed surface can't escape, is the main culprit in warming. The surface heats up and the heated surface emits more energy. Over time, it reaches equilibrium.

As an example, think of what happens to a car on a hot, sunny day. Sunlight floods in through the windows, heating the car's interior. Unable to get out, partly because the glass is opaque to it, the radiation from the heated interior stays inside and the car heats up. Adding to this effect is the fact that the car, like a greenhouse, is sealed off so no winds can get in to disperse the warm air.

For the time being, the equilibrium on Earth maintains our atmosphere at a comfortable and very habitable temperature. Without that, our atmosphere would be about 60 degrees F (33 degrees C) cooler, and our planet's average temperature would be a frigid 0 degrees F (-18 degrees C) instead of its balmy 60 degrees F (+15 degrees C).

Our atmosphere contains several "greenhouse gases" that prevent the infrared from escaping. The main ones are water vapor, carbon dioxide, nitrous oxide, and methane. As we mentioned, Mauna Loa has an important role in our understanding of the carbon dioxide in our atmosphere. Beginning around 1958, Charles Keeling at California's Scripps Institution of Oceanography started measuring the concentration of carbon dioxide from the summit of Mauna Loa. Beyond their great height, the location of Mauna Loa and Mauna Kea in the middle of the Pacific, free of local pollution, makes both outstanding sites for scientific observation.

When Keeling first began measuring, the concentration of carbon dioxide was roughly 310 parts per million. The amount undergoes a seasonal change every year. But apart from the seasonal blip, the "Keeling Curve," as it is known, shows an inexorable upward trend (Figure 4). The graph presents an alarming picture—and a compelling case for ensuring that we do not keep adding more carbon dioxide to our atmosphere.

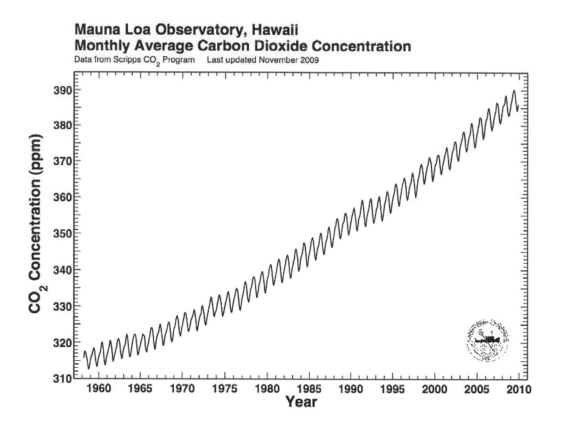

Mauna Loa Observatory, Hawaii
Monthly Average Carbon Dioxide Concentration
Data from Scripps CO_2 Program   Last updated November 2009

From the Scripps CO2 Program measured at the Mauna Loa Observatory, Hawaii, this is the Keeling Curve, monthly averages of the atmospheric carbon dioxide concentration over time. The concentration is given in parts per million in the mole fraction.

There are several ways that we humans are upping our atmosphere's carbon dioxide content. Most of the CO2 comes from burning fossil fuels; that is, coal, oil, and gas. Some of it comes from standard cars. Deforestation is a major culprit, adding a sizeable amount of carbon dioxide. The United Nations Climate Change Conference convenes every year to address the problem of carbon dioxide emissions, which has obvious political and economic implications. At the 2009 Conference, commonly called the Copenhagen Summit, the United States, China, India, South Africa, and Brazil drafted the Copenhagen Accord, formally recognizing climate change as a major global issue that warrants immediate action. The Conference was attended by representatives of 192 countries. Debates over the two-day conference produced serious disagreement and in the end, no legally binding agreements were reached.

The United Nations Climate Change Conference is held every year. At the 2012 Conference, held in Qatar, the nations agreed to extend the Kyoto Protocol, which was due to expire at the end of that year, to 2020. At the Copenhagen Conference, a major point of dispute was who should cover the cost of mitigating the impact of CO_2 emissions. At the 2012 Conference, the members agreed, at least in principle, that richer nations should be responsible for their failure to reduce greenhouse gases.

Non-government organizations (NGOs) are among the attendees at the United Nations Climate Change Conferences. In fact, many organizations have sprung up to combat global climate change. A group called 350.org is calling for the carbon dioxide concentration to be brought down to 350 ppm. (Remember, it was 310 ppm when Keeling began his work.) The bad news is that we have already failed in trying to keep it from going above 400.

The change in wording from "global warming" to the more encompassing "global climate change" recognizes that any warming would not occur equally in all parts of the world. Additionally, there is a substantial degree of local variations over time, and there are other factors that can cause transient cooling. Volcanoes, for example, cause temporary cooling.

Solar System Scoop

On May 9, 2013, the average CO_2 reading at Mauna Loa for 24 hours passed the symbolic, long-dreaded milestone of 400 ppm. The last time the air had that much carbon dioxide was three to five million years ago—way before humans existed. This time, there is no question about the role of human activity and what it means for the future of our beloved planet.

The carbon dioxide reading at Mauna Loa had been threatening to hit the 400 ppm mark for days. It was certainly no surprise when it finally did—and surpassed it. Scientists reacted quickly and emphatically. Pieter B. Tans, senior scientist at the Earth System Research Laboratory of the National Oceanic and Atmospheric Administration, declared that, "It symbolizes that so far we have failed miserably in tackling this problem." Ralph Keeling, who continues his father's legacy, at the Scripps Institution of Oceanography, was blunt about what the reading portends for our future. Said Keeling, "It means we are quickly losing the possibility of keeping the climate below what people thought were possibly tolerable thresholds."

As mentioned, carbon dioxide fluctuates on a seasonal basis. The level of CO_2 drops in summer when flourishing greenery draws carbon dioxide gas from the air. Carbon dioxide exceeding 400 ppm was initially seen in the Arctic in 2012. Mauna Loa had a few hourly readings surpassing the dreaded high. May 9th marked the first time the average level stayed at that point for an entire day. Scientists warn that even within a decade, there may be *no* carbon dioxide reading at any time, season, or any place on our Earth that falls below 400 ppm.

Regardless of what we call the phenomenon, unchecked warming would be disastrous. Even if our sea level rises by just a few feet (or meters), large areas in various parts of the world would be submerged, including the Maldives, a large part of Bangladesh, and coastal areas of the United States. The idea of parts of Manhattan under water is not just a science fiction plot—Hurricane Sandy showed us just that! Moreover, the effect is not necessary gradual or predictable. Masses of ice sliding into the ocean from Antarctica or Greenland could cause an abrupt rise in sea level.

The upward carbon dioxide curve is a significant icon for our time. It clearly illustrates that we need to make important changes to our energy and deforestation practices. Carbon dioxide has become the symbol of greenhouse gases (and rightly so). But there are other greenhouse gases, and though they exist in our atmosphere in much lower concentrations than CO2, molecule for molecule, some are actually more powerful. All signs point to the need to control the emission of greenhouse gases. And those signs were never as clear as the reading on Mauna Loa when carbon dioxide surpassed 400 ppm for the first time for a full day in eons.

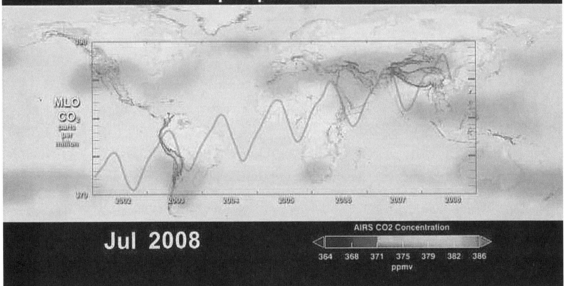

These carbon dioxide measurements were made at the Mauna Loa Observatory in Hawaii in a series started by Charles Keeling. The map in the background shows the amount of carbon dioxide in Earth's mid-troposphere in July 2008 as measured by NASA's Atmospheric Infrared Sounder on the Aqua spacecraft. Carbon dioxide continues to rise, along with its yearly cycle.

Light Pollution

As we discussed earlier, our ancestors could go out at night, and if the sky was clear they would have a magnificent view of the stars and planets. That changed after electricity was invented, lighting up cities and towns. Today, most of us live in areas that are inundated with light. In urban areas especially, the light scatters across the sky, blocking most stars and all but the brightest planets from view.

The term "light pollution" arose from recognition that the light shining upward is just what that term implies: a type of pollution by light. How many times have you looked up at the sky and wished you could see all the stars that appear on a sky map? The constellations were clear to the ancients who named them. Light pollution keeps many (or most) people from enjoying them. Any light that extends up in the sky is generally wasted light. If you put a reflector on top of a streetlight, the light is directed downward, illuminating the ground. The safest and most efficient lighting illuminates the ground evenly without contrast or light misdirected. LED lighting may offer a prototype for the streetlights of the future—lighting the way for pedestrians and drivers *without* throwing light into the sky.

Solar System Scoop

A team of researchers from Mexico and Japan came up with a lighting design to combat light pollution. In the innovative design, street lamps would be equipped with a cluster of LEDs and specialized lenses placed inside a reflective box covered by a microlens based on the principles of an insect's eye. The light would only shine where it is needed.

For drivers and pedestrians, the LED street lights would minimize glare and maximize visual acuity. As an added bonus, they would also be more energy efficient. And for stargazers everywhere—especially in metropolitan areas—the skies would have far less light pollution and many more visible shining stars. The catch: a working prototype for the LED street lamps is still in the future. So far, it exists just as a great idea, though the researchers assure us, a viable one.

Satellite maps of the sky at night present us with interesting view of our planet, from the perspective of artificial and natural light. We can see cities, highways, train routes, bridges, and other features that are lit up at night. The light areas contrast with the darkness covering most of the oceans, big forests, deserts, and even some less prosperous cities and countries.

The Earth at night, looking down from space. There are concentrations of light in some places, the most urbanized, and large regions of continents where there is little light at night.

Solar System Scoop

As an example of how light pollution affects our view of the night sky, in 1994 when the Northridge earthquake produced a power outage in Los Angeles, local emergency centers were inundated with calls from nervous residents who saw a "giant, silvery cloud" in the dark night sky. The ominous cloud was the Milky Way! Obscured by the bright lights of L.A., no one had ever seen it before. Noting that many callers refused to believe that the strange cloud they saw in their darkened city was actually the real night sky, Ed Krupp, the director of the Griffith Observatory, commented, "Since so many of us never see a non-light-polluted night sky from one year to the next, a mythology about what the people *think* a true star-filled sky looks like has emerged."

Tidal Forces, the Moon, and the Sun

The power of tidal forces is felt throughout the universe. The relationship of the tides to the Moon is well-known, but they have other awesome effects such as Saturn's beautiful rings.

The power of tidal forces is felt throughout the universe. The relationship of the tides to the Moon is well-known, but they have other awesome effects such as Saturn's beautiful rings.

And though we tend to associate the tides with the Moon, the tides we experience on Earth are the combined effect of the gravitational forces of the Moon and the Sun.

Ocean water cyclicaly rushing into and away from shore and sometimes reaching amazing heights before it recedes are the obvious manifestation of tidal forces on Earth. These "tidal bulges" result from a distortion of water and terrestrial material produced by the gravitational forces of the Moon and the Sun and the rotation of the Earth.

The defining property of a tidal force is that it is a *differential* force; that is, the difference of two forces. Variations in the gravitational forces as they act on the materials our Earth is made up of (and we live on a very watery planet) determine the extent of the tidal bulge. On the side of the Earth closest the Moon, the Moon's gravitational pull on the ocean a is more powerful than its pull on the more massive solid ground. On the other hand, because the force of gravity is weakened by distance, the Moon's gravitational pull on the solid terrain is stronger than its pull on the ocean point on the side of the Earth farthest away from the Moon.

Though the Sun's gravity is certainly stronger than Earth's, it is much farther away from us than the Moon. As a result, there is less variation in its gravitational pull from the near side to the far side. The solar tide is a lot weaker than the lunar tide (by a factor of about 4).

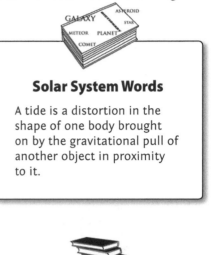

Solar System Words

A tide is a distortion in the shape of one body brought on by the gravitational pull of another object in proximity to it.

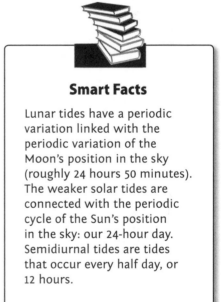

Smart Facts

Lunar tides have a periodic variation linked with the periodic variation of the Moon's position in the sky (roughly 24 hours 50 minutes). The weaker solar tides are connected with the periodic cycle of the Sun's position in the sky: our 24-hour day. Semidiurnal tides are tides that occur every half day, or 12 hours.

The actual tides at any given location are also subject to local geological conditions such as the shape of the coastline, the slope of the ocean floor, and ocean currents. Water moves around in ocean basins and bays. Anyone who has visited a lot of coastal areas is familiar with the many differences in the tides.

Each month when the Sun and the Moon are aligned at a new moon or a full moon, the rising tide is so strong that it seems to "spring up." This phenomenon has earned the month's highest tide the name "spring tide." The lowest tides occur when the Sun and the Moon are at right angles to each other from our vantage point on Earth (and we see a first-quarter or third-quarter moon). These low tides are called "neap tides."

The Earth's Magnetic Field

If you look at a compass, you can see that the needle tends to align in a north-south direction. The Earth has a magnetic field similar to a bar magnet's (though more complicated) with one pole located in Hudson Bay, Canada (not at the North Pole). The magnetic field protects our planet from incoming charged particles launched from the Sun and other parts of the universe.

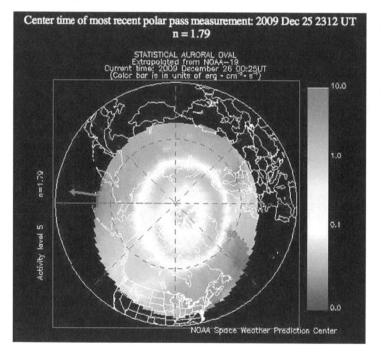

The northern-hemisphere auroral oval centered over Hudson Bay from NOAA's Space Weather Prediction Center.

The Van Allen belts range from a few hundred miles high to about eight times the radius of the Earth. Since Van Allen's discovery, NASA has sent numerous satellites into orbit to explore the Van Allen belts and the Earth's magnetic field.

Solar System Scoop

James Van Allen is credited with making the first discovery of the protective nature of Earth's magnetic field using data from an orbiting satellite. In 1958, Van Allen headed a project measuring charged particles in space with Explorer 1. He found rings surrounding the Earth filled with orbiting charged particles. As we discussed earlier, these rings were named the Van Allen belts. Some of the outer planets have even stronger magnetic fields and more far-reaching versions of the Van Allen belts found around Earth. The Sun is the source of these particles, which ultimately impact on planetary magnetic fields forming the belts around all planets with a magnetic field. Without the protective magnetic field, these particles impact the planetary surface directly.

Observations of the auroras play an important role in our knowledge of the Earth's magnetic field. Through images from space, we can see the auroral ovals around the north and south magnetic poles. Solar storms fill the auroral ovals with particles, sometimes causing them to expand toward the equator. This gives more people a chance to see the elusive auroras from Earth.

CHAPTER 7

The Moon: In Earth's Gravity and Lighting the Night Sky

In This Chapter

➤ The Moon's phases

➤ The lunar landscape

➤ Lunar eclipses

➤ The Apollo missions: moonwalks and moon rocks

➤ Robotic missions: a global enterprise

The term "lunar" comes from Luna, the ancient Roman Moon deity (Selene in Greek mythology). Her symbol was a crescent Moon, and she was shown driving a two-yoked chariot with one black horse and one white horse to signify that the Moon is visible night and day. And indeed, if the sky is clear enough, we can see the Moon in the daytime as well as at night. At night, the Moon, and especially the full Moon, dominates the dark sky. The Moon's changing phases add to its mystique as one of our most fascinating celestial objects.

Scientific interest in the Moon has endowed it with a particularly unique distinction. Other than our own planet, the Moon is the only celestial body where humans have ever set foot. There have been no humans walking across the lunar landscape for four decades, but the Moon will never lose its allure.

The Moon takes up ("subtends" to astronomers) a half degree in the sky. (For comparison, if you stretch out your arm, your thumb subtends about 2 degrees.) The Moon's angular size makes it far bigger than any other celestial bodies we can see—it even dwarfs the field of view of a large telescope. It is spectacular viewed through binoculars or a small telescope—great news for amateur astronomers!

The Moon and other objects in the sky in their true comparative angular sizes.

The crater Tycho, with white rays coming out from it, is at center bottom.

The crater Copernicus is near the equator midway across on the left side, with rays of material expelled by the impact that made it. At the bottom there are deep-sky celestial objects imaged with the Hubble Space Telescope.

The Lunar Phases

Our natural satellite completes its orbit around the Earth every 27½ days. That interval refers to its "sidereal period," the length of its orbit in relation to the stars, or if we were high up looking down. From our earthly perspective, the Moon goes through a full set of phases every 29½ days. What accounts for the two-day difference? By the time the Moon orbits the Earth, our planet has made about 1/12th of its journey around the Sun. The Moon has to keep moving along on its orbit for almost two more days before it catches up with the Earth-Sun line, which begins the next set of phases.

The Moon's orbit around the Earth is tipped roughly 5 degrees from the Earth's orbit around the Sun. Because of that, when the Moon comes between the Earth and the Sun, it is not always exact. The Moon is usually a bit higher or lower than the Sun. The time when the Moon and Sun are closest is the "new Moon." During that phase, only the far side of the Moon is illuminated and all we can see from Earth is its dark side.

Solar System Scoop

The side of the Moon we see as it orbits the Earth is always the same. Over billions of years, a slight bulge in mass built up on that side, and the Earth's gravity keeps it locked facing toward us. So what we see from Earth is always the "near side"; we can only see the "far side" from space.

Depending on the Moon's position in its orbit around the Earth, any half of the Moon can be sunlit. Sometimes we can see the entire illuminated area, sometimes none of it, and most of the time, some of it. At any time, the "dark side" of the Moon is the side facing away from the Sun, and we can usually see at least part of it.

The "dark side" of the Moon should not be confused with the "far side."

When the Moon and the Sun are in precise alignment, we are treated to the event of solar eclipse. Total solar eclipses are rare. The last total solar eclipse was on November 13, 2012, and there will not be another one until 2015. But on November 3, 2013, we can expect an even more unusual event—a hybrid or annular/total eclipse in which people in certain parts of the world will see an annular eclipse and in other areas, they'll see a total eclipse. We'll discuss solar eclipses in the chapter on the Sun.

A slim crescent begins to appear a day or two after the new Moon, when the Moon has moved enough in its orbit so we can see a bit of the side that is lit. We see the crescent phase for about a week.

By the end of the week that the Moon appears as a crescent, it has gone a quarter of the way through its phases. Phase-wise, that means it is a first-quarter Moon. Visually, though, half the Moon is lit, which means it is also a half Moon.

For the next week, the Moon is in the gibbous phase, where more than half the face we see is illuminated. The somewhat strange-sounding term means "hump" in Latin.

Solar System Scoop

Leonardo Da Vinci was unquestionably brilliant and imaginative. In the 16th century, he explained the phenomenon of "earthshine," which has fascinated and puzzled people for millennia. Looking like the ghost of a full Moon nestled in the crescent Moon, earthshine has been called "the old Moon in the new Moon's arms." Leonardo realized that the ghostly light was caused by sunlight reflecting off the Earth.

As the lighted portion grows bigger, we say that the Moon is waxing. The three partial phases we see are the waxing crescent, first quarter, and waxing gibbous Moon. Finally, the next week, the Moon is completely illuminated and (if the sky is clear enough) we can enjoy the sight of a gleaming full Moon. During the full Moon, the Sun is roughly behind the Earth. If it were directly behind us, the Earth's shadow would hit the Moon, eclipsing it from our view. We'll get to that later. In most months, the Earth's shadow passes above or below the Moon, so the full Moon is fully visible.

After the full Moon, the gibbous Moon reappears, but this time it is a waning gibbous, lasting about a week. The third-quarter Moon—and second half Moon, as we see it—follows the waning gibbous. The waning crescents reappear for the last set of phases, and the full cycle completed, it's time for a new Moon again.

Following the Moon's Phases

The Moon's phases follow a pattern that enables us to tell the time of the month. Close to the Sun in the sky, the new Moon rises and sets with the Sun. As moonrise occurs roughly 50 minutes later each day, about two days after the new Moon, a narrow sliver appears and sets close to two hours following sunset. It takes two days after the new Moon to see the crescent because a one-day-old Moon is so thin and faint that it disappears in the twilight.

The first quarter Moon shines halfway up in the sky at sunset and sets around midnight, about six hours later. The bright, gleaming full Moon lights up the sky nearly all night long. Following the full Moon, the waning gibbous rises at a later time each night. By the time the third quarter Moon comes around, it rises at about midnight, placing it high in the sky at sunrise. Given this timing, the waxing gibbous Moon appears to a much smaller audience than the waxing crescent that shines in the evening sky.

Lunar Librations

Due to the Earth's gravitational lock on the Moon, its far side is hidden from our terrestrial view. It was not until the lunar space missions that we could see the Moon's rocky and heavily cratered far side. At the same time, it is not entirely true that we can only see half the Moon. As a result of lunar librations, we actually see slightly more than half the Moon from Earth. The combined effects of the angle of the Moon's orbit around the Earth, and the fact that its speed as it journeys along its elliptical orbit varies according to its distance from Earth (Kepler's second law), allows us glimpses of the far side, and at some points glimpses over the top or the bottom. The result is that about 5/8 of the Moon is visible at different times. The variations that occur in our angle of view are known as librations.

Interesting animations of the Moon's librations can be found at:

http://antwrp.gsfc.nasa.gov/apod/image/0709/lunation_ajc.gif
http://antwrp.gsfc.nasa.gov/apod/ap051113.html

In the second simulation, you can see that Mare Crisium, close to the right-hand limb, sometimes appears right on the edge and sometimes farther inside the lunar disk.

The Craggy Moon

As we discussed in Chapter 2, Galileo aimed his telescope at the Moon in 1609. With the telescope's small field of view, he could only see part of the Moon at one time, but as an expert Renaissance draftsman, he knew he was seeing shadows. He labeled the smooth, expansive areas he saw *maria,* from the Latin word *mare,* for sea. He distinguished mountains and measured their height, and picked out the craters—the Moon's most famous features.

Galileo presented his discoveries as engravings in his 1610 work *Sidereus Nuncius (The Starry Messenger).* Some of the features might have been exaggerated to make more of an impact (whether it was the engraver's idea or Galileo's, we have no way of knowing). But embellished or not, Galileo was the first to discover, and to illustrate for the world, the features that mark the surface of the mysterious Moon. Since Galileo, there have been numerous maps of the Moon with varying degrees of accuracy. Today, we have high-resolution images taken from lunar orbit and Earth to show us what the Moon's surface really looks like. We also have some excellent photographs of the lunar landscape taken by the Apollo astronauts and showing close-ups of craters.

A view of the Moon's far side, in a photograph taken by the Apollo 16A mission's astronauts.

Almost all of the far side is covered with craters. The Earth did not protect it from being clobbered by meteorites billions of years ago, so it has deep and extensive impact craters. Adding to the cratered appearance of the far side of the Moon, it did not experience the eruptions of lava that smoothed out much of the landscape of the near side.

A photograph of the craters Eratosthenes and Copernicus, and the region near them, taken by astronauts in orbit in Apollo 17 in 1972.

In Figure 1 in this chapter, we can clearly see the *maria*. The largest sea, on the right side of the image, is Mare Tranquillitatis (Sea of Tranquility), the site of the first lunar landing. To its left is Mare Serenitatis (Sea of Serenity). The oval-shaped sea on the far right is Mare Crisium (Sea of Crisis). On the left side, Mare Imbrium (Sea of Showers/Rain) is above the center and just above the crater Copernicus. At the extreme upper left is Oceanus Procellarum (Ocean of Storms).

Eclipses of the Moon

A lunar eclipse occurs when the Earth's shadow falls upon the Moon, obscuring it from our view. The Moon's orbit is tilted about 5 degrees in relation to the Earth's path around the Sun; a lunar eclipse is the event that takes place when the two orbits cross from our terrestrial vantage point.

Buzz Aldrin on the Moon, in a historic photograph from the Apollo 11 mission, taken by Neil Armstrong. Armstrong and Aldrin left a plaque on the lunar surface reading, "Here men from the planet Earth first set foot upon the Moon July 1969 A.D. We came in peace for all mankind."

When the Moon is completely engulfed by the Earth's shadow, all that illuminates it is a muted red light that bends around the edges of the Earth in our atmosphere. Consequently, what we see during a total lunar eclipse is a faint, reddish object. Totality during a lunar eclipse lasts approximately one hour and can be seen anywhere in the world where the Moon is visible. That event is dramatically different from a total solar eclipse, which is visible only within a narrow strip about 200 miles (322 kilometers) in diameter and thousands of miles long.

The last total lunar eclipse occurred on December 10, 2012. Unfortunately, the next total eclipse will not take place until April 15, 2014. It will be visible in the Americas, Australia, and the Pacific.

Walking on the Moon

The idea of sending astronauts to the Moon emerged as an antidote to the problems faced by the U.S. in the Cold War. The disconcerting lag of the U.S. in the space race was underscored on April 12, 1961, when Soviet cosmonaut Yuri Gagarin became the first person to orbit the Earth. Later that month, the failed Bay of Pigs invasion of Cuba intensified pressure on President John F. Kennedy to show that the U.S. could not only keep up with, but also surpass the Soviets. Kennedy consulted with Vice President Lyndon Johnson, who proposed sending astronauts to the Moon. After discussing the plan with NASA Administrator James Webb as well as other officials, Kennedy concluded that sending Americans to land on the Moon was a technologically challenging venture, but it was also the area of space exploration in which the U.S. had a potential advantage. On May 25, 1961, in a dramatic and stirring speech, Kennedy announced that the U.S. would send men to the Moon and bring them back safely to Earth by the end of the decade.

Achieving that ambitious goal is still recognized as one of the most intensive enterprises of recent history. NASA personnel involved in the lunar project were under phenomenal pressure. On July 20, 1969, the world watched astronauts Neil Armstrong and Buzz Aldrin land on the barren lunar landscape. They landed in Mare Tranquillitatis and named their site Tranquility Base. Stepping out of the space capsule in their bulky spacesuits, Armstrong and Aldrin bounded across the lunar surface under the Moon's weak gravity (only 1/6 of our gravity here on Earth). After a few hours, they rendezvoused with fellow astronaut Michael Collins, who remained in the lunar orbiter. As we know, the mission was carried out safely, as President Kennedy promised. The now legendary astronauts came safely back to Earth to a hero's welcome.

Armstrong, Aldrin, and their successors all left equipment on the Moon to send measurements back to Earth. Seismographs, for example, recorded moonquakes. The famous "moon rocks," lunar samples brought back by Aldrin and Armstrong, have been analyzed extensively ever since. Today, we have more than half a ton of moon rocks to study: 840 pounds (382 kilograms) of rocks were brought back to Earth by 12 astronauts over six Apollo missions.

Harrison Schmitt, the only credentialed scientist to have been an astronaut on the Moon, stands next to a huge, split lunar boulder during his Apollo 17 expedition in 1972.

Moon Rocks and What They Reveal

Once brought back to Earth, the rocks were taken to laboratories such as the one at the Johnson Space Center (home of the rock depository) for intensive examination. Analysis of the rocks shows they are not quite as exotic as the term "moon rock" implies. The types of rocks dotting the Moon are actually very similar to the rocks here on Earth. Almost all of them are igneous rocks, the type of rock formed by cooling lava. Most of the *maria* are composed of basalts, which are common igneous rocks. Some of the astronauts' finds are breccias, rocks created when other rocks are broken apart and reformed. Remember, the Moon was relentlessly bombarded by meteorites.

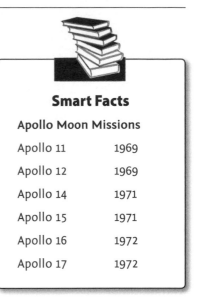

Smart Facts

Apollo Moon Missions

Apollo 11	1969
Apollo 12	1969
Apollo 14	1971
Apollo 15	1971
Apollo 16	1972
Apollo 17	1972

Along with scrutinizing their composition, the geologists dated the ages of the Moon rocks. Some of the oldest rocks date back 4.44 billion years. The younger rocks date back 3.1 billion years. The Moon's activity ended at that point.

In between the times of the youngest and oldest rocks was an era where lava flowed over the lunar surface, smoothing parts of the craggy landscape. The *maria* were produced by that smoothing effect. For example, the rocks Armstrong and Aldrin retrieved from the Sea of Tranquility were dated to 3.6 billion years ago.

With a wealth of samples from the Apollo missions and some unmanned Russian moon missions, scientists have constructed a lunar chronology. The original lunar surface, after the Moon formed 4.5 million years ago, was warm. Heat from the Moon's formation, intensive bombardment by meteorites, or internal radioactivity could have all played a role in keeping the surface temperature up. Roughly the top 300 miles (500 kilometers) were molten, cooling down within about 0.2 billion years. Unremitting bombardment between 4.2 and 3.9 billion years ago produced most of the Moon's famous craters. Radioactive energy probably caused the internal flow, which started about 3.8 million years ago and created the major *maria*. The Moon has changed very minimally in 3.1 billion years. There was some lava flow 2.5 million years ago, but the Moon we see today is essentially the same as it was at that point in primordial history.

The Moon's unchanging landscape provides us with important clues to what was happening in the early solar system. On our own planet, plate tectonics and weather have obliterated almost all evidence of the original surface and earliest rocks. Fortunately, we can study the Moon to discover what was going on in that chaotic early solar system.

Earthrise over the lunar surface, from the Japanese space agency's Kaguya spacecraft in 2008.

After Apollo

July 20, 2009, marked the 40th anniversary of the first lunar landing. This time, the historic event was celebrated over the world online and through social media, attesting to the major technological advances over the last decades. It is amazing that we were able to send people to the Moon and safely back 40 years ago—and equally amazing that with ever more sophisticated technologies there have been no human Moon missions since. In 1972, Apollo 17 carried Gene Cernan and geologist Harrison Schmidt to the Moon. That was the last of the Apollo missions. NASA's first frenzied effort to land people on the Moon took less than nine years.

At the present time, NASA has no plans to return to the Moon. The ultimate goal for human space travel is Mars, but that lies in the distant future.

If astronauts have not been hopping across the lunar surface, the Moon always has been and continues to be a popular site for space exploration. Overlapping with the Apollo missions in 1969 and 1970, the Soviet Union sent two unmanned spacecraft to bring back soil samples. In 1976, they sent a third spacecraft, Luna 24. The three lunar missions produced 0.3 kilograms of lunar soil, some of it drilled a half meter below the Moon's surface. Two additional Soviet missions carried rovers, Lunokhod 1 and Lunokhod 2. For several months in 1970 and 1973, respectively, they traveled over the Moon's surface. The first images of the Moon's far side were sent from a Soviet spacecraft. We could actually credit the accumulation of Soviet "firsts" with spurring the U.S. effort to send the first humans to the Moon. With the success of Apollo 11, the Soviets turned their interests away from the race to the Moon.

In the 1990s, a few missions flew by the Moon or entered its orbit. Japan sent a small spacecraft into lunar orbit in 1990. In 1991, NASA's Galileo mission flew past the Moon and tested some of its cameras en route to Jupiter. In 1994, a joint venture by the Ballistic Missile Defense Organization (BMDO) and NASA sent Clementine (officially the Deep Space Program Science Experiment) into orbit around the Moon for 71 days to test image cameras and other spacecraft components under conditions of prolonged space exposure and to map the complete lunar surface. Clementine had a major advantage over the Apollo spacecraft, as it flew in polar orbit, capturing all areas of the Moon, while the Apollo missions were limited to low lunar altitudes.

Solar System Words

If you're wondering how Clementine got its name, in the old song "Oh My Darling, Clementine," Clementine was a prospector's daughter, and the aim of the Clementine mission was to map the distribution of minerals. Clementine in the song was also "lost and gone forever," as the spacecraft would be after its mission was over.

In 1998, four years after Clementine, NASA launched a Lunar Prospector mission. One of its tasks was to search near the lunar poles for any ice that might still be lurking inside the interior of a perpetually shadowed crater. In 1999, Lunar Prospector was crashed into a polar crater, on the exciting theory that it might eject a plume of water vapor that would be captured by telescopes on Earth. Unfortunately, the mission did not produce its intended effect. No water vapor was found, but the idea was certainly not abandoned. Controlled impact experiments with the goal of detecting water vapor in the gas and dust released by the crash are an ongoing part of lunar exploration.

In 2003, the European Space Agency sent a test satellite to the Moon as part of their Small Mission for Advanced Research and Technology (SMART). The satellite was propelled by the ejection of ions instead of the chemical rockets used by all of the other Moon missions. It was a great success! The satellite transmitted numerous images and other data before it was crashed into the Moon in 2006.

Hiten (Flying Angel) was the name of the 1990 Japanese lunar mission. It orbited the Moon before being crashed into the lunar surface in 1993. In 2003, the Japan Aerospace Exploration Agency (JAXA) was formed as a merger of three aerospace organizations: the Institute of Space and Astronautical Science, the National Aerospace Laboratory of Japan, and the National Space Development Agency of Japan. In 2007, JAXA announced plans for its **Sel**enographic and **En**gineering **E**xplorer (SELENE) mission, nicknamed Kaguya. (Remember, Selene is the Greek name for the Moon goddess; Kaguya was a Moon princess in a traditional Japanese story and was chosen by the public.) Kaguya was launched on September 14, 2007. After being in lunar orbit, SELENE crashed into the lunar surface in 2009, sending back spectacular videos of the event.

NASA's LCROSS rocket crashes into the Moon on October 9, 2009, photographed from the trailing spacecraft. All that was seen was the faint puff of light in the middle, enlarged at lower left.

The Chinese Lunar Exploration Program (also called the Chang'e Program) of the Chinese National Space Administration launched its first lunar orbiter on October 24, 2007. The successful launch was followed by another successful mission on October 1, 2010. These two robotic missions represent the first phase of the Chang'e Program, which ultimately plans to send humans to land on the moon, though not until 2025 or later. In the interim, Chang'e 3 is scheduled to send two lunar landers equipped with rovers in 2014, followed in 2017 by Chang'e 4, a sample return mission. Chang'e 4 is expected to bring back up to two kilograms of lunar samples.

In October 2008, the Indian Space Research Organization launched Chandrayaan-1. (*Chandrayaan* means Moon Traveler in Sanskrit, after the Hindu moon god, Chandra.) Similar to the Lunar Prospector and a few other spacecraft, Chandrayaan sent a probe to crash into the Moon's surface near the South Pole with the aim of sending up a plume that could be analyzed for water ice. Eventually besieged by technical problems, the orbiter stopped transmitting signals in August 2009. Though the mission was cut short, it succeeded in achieving most of its goals, including sending back the lunar soil to be analyzed for water ice. Onboard, Chandrayaan-1 carried a global assortment of experiments, from the U.S., the U.K., the European Space Agency, Bulgaria, Germany, Norway, and Poland.

In June 2009, NASA launched its Lunar Reconnaissance Orbiter (LRO), which is still traveling in polar orbit around the Moon. From a low orbit, only 50 kilometers high, it is carrying out a detailed mapping mission. The same launch also carried the **L**unar **Cr**ater **O**bservation and **S**ensing **S**atellite (LCROSS) on another quest for water ice in a shadowed crater near the lunar pole. LCROSS crashed into the crater on October 9, 2009. First the rocket's upper stage crashed and the probe that followed it sent back images. Telescopes were set up all over the world and anticipation was certainly high, but neither images nor spectroscopy showed a plume spurting up from the crater. All that was seen through the telescopes was a faint wisp of light.

Analysis seems to indicate that there was some presence of water. An exciting discovery, but whether there is enough to be useful to future astronauts is a question for debate.

In March 2013, NASA released images taken by LRO of the impact of the twin **G**ravity **R**ecovery and **I**nterior **L**aboratory (GRAIL) spacecraft as they crashed the Moon's surface near the North Pole. The GRAIL spacecraft were sent into lunar orbit in 2011 to map the Moon's gravitational field in meticulous detail and provide new insight into its interior. GRAIL succeeded in mapping the Moon's gravitational field in unparalleled detail.

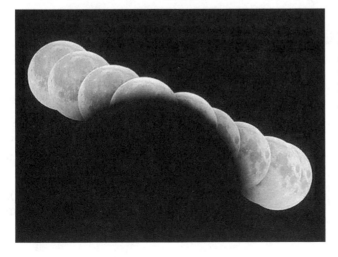

A partial eclipse of the Moon, with the position of the Earth's shadow held steady in the frame.
Credit: Anthony Ayiomamitis; http://www.perseus.gr/Downloads /eclipse-lunar-2008-08-16-umbra.zip)

NASA's next lunar venture, the **L**unar **A**tmosphere and **D**ust Environment Explorer (LADEE) is scheduled to be launched in 2013 to orbit the Moon's equator. The small spacecraft's mission is evident in its title: LADEE will examine the Moon's very thin atmosphere (exosphere) and dust in its vicinity.

Luna-Glob (lunar sphere), the lunar exploration program of the Russian Federal Space Agency was planned as early as 1997, but it was fraught with continual setbacks and only revived in the last few years. Luna-Glob 1 was originally scheduled for launch in 2012. Its launch date has since been changed to 2015. Gathering data on seismic activity is a major goal of the mission.

It's very clear that all the current and planned lunar missions have one thing in common: they all carry robotic instruments. The only planned expedition carrying humans (that we know of) is Chang'e 5, and that is tenuous and not expected until 2025 or 2030.

CHAPTER 8

Mars: The Red Planet Beckons

In This Chapter

➤ Observing Mars from Earth

➤ Mars from space: soaring peaks and huge canyons

➤ Mars's atmosphere

➤ Landing on Mars and roving around

➤ Life on Mars?

Of all the planets in our solar system, none has held as much fascination for humans as Mars. Named for the Roman god of war for its blood red color, Mars gleams brightly in the night sky, often outshining the stars. And unlike the toxic Venus, Mars has always seemed to hold the possibility that someday we humans might actually be able to live there. People have wondered what Mars is really like for millennia, and finally, we have the technology to find out. NASA has even set its sights on a crewed expedition to Mars. The big question is *when?*

As it rises and sets from day to day, Mars travels across the sky slightly faster than the stars…most of the time. At some points, it slows and even seems to move backward with respect to the stars. (On some occasions it even seems to move a bit sideways, making a loop over time.) The planet's retrograde motion led Copernicus to recognize that the Sun is the center of our solar system. Kepler determined that planetary orbits are elliptical by studying Tycho Brahe's meticulous notations of Mars's movement among the stars. Only Mercury has a more eccentric orbit than Mars.

And only Mercury of the planets in our solar system is smaller than Mars. In the nineteenth century, when Mars was a very popular target for telescopes, its disk appeared small to even the cutting-edge telescopes of the day. Some of its features were visible, and given the planet's apparent similarity to Earth, they were not necessarily interpreted accurately. Similar to Earth's, Mars's axis of rotation is tipped with respect to the ecliptic (the plane of Earth's orbit around the sky), meaning that Mars has seasons. During the Martian springtime, the telescopic images showed dark patches spreading out over the Martian landscape, looking like vegetation. We now know that the "vegetation" is really the reddish dust blown around by seasonal winds. Without sophisticated instruments, the idea that there was plant life on Mars could not really be dispelled, so it persisted for decades.

Furthermore, the idea of life on Mars has always been enticing. In 1877, the Italian astronomer Giovanni Schiaparelli recorded that he observed "canali" on Mars. What he really meant was the Italian word for "channels." In English it was translated into "canals"— *artificial* waterways. Someone had to have dug those canals! The implication was that there were Martians digging canals, possibly as irrigation for those dark patches of "vegetation" that seemed to appear in the spring. Percival Lowell, an American astronomer and mathematician, was the main proponent of the theory that there were canals on Mars. In the late nineteenth century, he published numerous books with intricate illustrations of what he supposedly saw, with detailed descriptions of the apparent Martian-created canals. Lowell's ideas were popular with the general public, but derided by astronomers.

Solar System Scoop

If Martians were capable of building canals, they also might be capable of traveling to other planets. H.G. Wells built on that theme in his 1898 science fiction novel *The War of the Worlds*. Living up to being named for the god of war, the Martians invaded England. On the night before Halloween 1938, Orson Welles directed and narrated a radio broadcast of *The War of the Worlds*, set as a Martian invasion of New Jersey. Welles's adapted version cleverly (perhaps too cleverly) made use of news broadcast techniques—which is what many Americans thought they were listening to. There were newspaper accounts of "mass panic" in response to the purported invasion; they were probably exaggerated, but there was no question that many radio listeners were frightened by what they thought was a real invasion from Mars.

In a real historic event, in February 2013, NASA's Curiosity rover drilled into the Martian surface, dislodging powdery rock samples. Scientists analyzing the precious find speculate that there might once have been life on Mars—billions of years ago, that is, and the life forms would have been microbes. However, no life forms have yet been found, although it is clear that liquid water once flowed across Mars' surface. Mars will always be an exciting place, and we have barely begun to discover what the red planet is really like. Images taken by the Hubble Space Telescope and spacecraft orbiting Mars or crossing its surface reveal a wondrous planet that deserves the attention it has gotten through history. If tiny microbes are far from the Martians of our science fiction fantasies, at least they are not capable of attacking Earth.

Figure 1. Four views of Mars, taken within a day with the Hubble Space Telescope as it rotates. The giant volcanoes on the Tharsis ridge are at lower left of the upper-right image, with Olympus Mons extending through the clouds, and on the right side of the lower-left image. At lower right, there is the dark region Syrtis Major. The bright feature near Mars's south limb is Hellas, an impact basin that is 1,200 miles across.

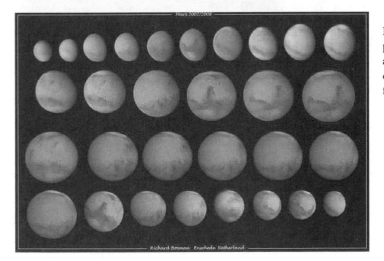

Figure 2. A series of photographs taken over about a year showing changing sizes of Mars as seen from Earth.

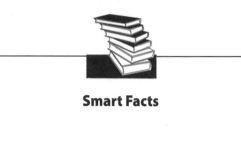

Smart Facts

Mars Facts

Mars's orbit

Average distance from Sun	1.52 AU	228 million kilometers
Period by the stars	1.88 years = 687 days	
Period from the Earth	780 days	
Orbit's eccentricity	0.09	
Orbit's tilt	1.9°	

Mars the planet

Mars's diameter	6,794 km	0.53 Earth's
Mars's mass	11% Earth's	
Mars's density	3.94 times that of water	
Mars's surface gravity	0.38 Earth's	
Sidereal rotation period	24 hours 37 minutes	

Solar System Scoop

Mars shines steadily in the sky, a pinkish or reddish object that rivals all the stars, with the exception of Sirius, in brightness. Ever the attraction for amateur astronomers, its details are hard to distinguish, even with a fairly large telescope. What you'll see through a telescope lens is primarily the reddish disk. You can see the darker regions rotating with the planets, and the polar caps waxing and waning with the Martian seasons.

Mars is in opposition—on the opposite side of the Earth from the Sun—about every 26 months. At those points, it rises at sunset and can be seen in the sky all night. The next opposition dates are April 8, 2014, and May 22, 2016.

Smart Facts

Here are some of the missions to our ruddy neighbor (NASA, unless otherwise specified):

Mariner 4, 1965	first flyby with photographs
Mariner 9, 1971	first orbiter, waited out a dust storm
Viking 1 and 2, 1976	landers
Mars Pathfinder, 1997	Sojourner lander (now the Carl Sagan Memorial Station)
Mars Global Surveyor, 1997–2006	
Mars Odyssey, 2001–	

Mars Express (European Space Agency), 2003– Beagle 2 lander failed
Mars Reconnaissance Orbiter, 2006–
Mars Exploration Rover: Spirit, 2004–
Mars Exploration Rover: Opportunity, 2004–

Phoenix lander, 2008	lander, searching for microbial-safe environmentsMars
Science Laboratory, 2011	landed the Curiosity rover

Mars Atmosphere and Volatile EvolutioN (MAVEN), scheduled for late 2013 to study the Martian atmosphere

Mars from Orbit

As part of the ongoing exploration of Mars, spacecraft from NASA and the European Space Agency (ESA) are making their way around Mars, sending back finely detailed images of the reddish surface. A primary goal of the missions to Mars is discovering the areas that are the safest and most interesting (the two don't always match) for robotic landers and ultimately, for human visitors. The spacecraft are usually launched when Mars and Earth are at their closest proximity every 26 months to shorten the transit time.

Robotic orbiters have been sending us pictures of Mars for more than forty years. We now have a comprehensive and detailed image of the planet's entire surface. The Martian atmosphere is very thin, only about 1/100 the density of our own atmosphere. The thin atmosphere is a great advantage for space cameras—they get an excellent view of the surface.

At the same time, while Mars is not obscured by thick clouds like Venus, it is subject to dust storms that have the same effect in blocking its surface from view. That happened in 1971 when the first orbiter, NASA's Mariner 9, arrived at the inopportune time of a global dust storm. It was months before the dust settled, gradually revealing first one high volcano and then several others. We now know that Mars has a vast high plateau, called Tharsis (though "Tharsis" sounds like the name of a Greek deity, it refers to a large, elevated region). Near

Tharsis is Olympus Mons, an enormous high volcano. (In this case, the name does come from Greek deities; Olympus was the home of the ancient gods.) Even before spacecraft traveled to Mars, Olympus was visible as a bright spot on images taken from Earth. It was given the name Nix Olympica, the snow of Olympus.

Olympus Mons is humongous. Its base stretches 600 kilometers across, and it soars to about 21 kilometers in height (truly Olympian proportions!). The vast crater atop the volcano could engulf all of Manhattan Island so that not even the top of the Empire State building would show over the rim. Mount Everest measured from sea level and Mauna Kea measured up from the ocean floor would both be dwarfed by Olympus Mons.

Mauna Kea owes its size to a hot spot of volcanic activity deep under the ocean. As a result of shifts in a continental plate, different points of land have been over the spurting lava plume, forming the Hawaiian Island chain. Mars has no continental drift. Consequently, lava built up from a similar plume would be concentrated in one place. That would probably account for the enormous size of Olympus Mons.

The Mars orbiters also revealed a chain of canyons, roughly the width of the continental United States (3,000 miles/5,000 kilometers long). Discovered by the Mariner spacecraft, they were named Valles Marineris. This Martian feature is similar in scale to the huge Rift Valley that runs north-and-south through Africa.

The orbiters have also captured many smaller channels on the Martian terrain. As it turns out, Schiaparelli was right in that there are channels on Mars—but not in the locations he mapped. We now know that what Schiaparelli and Lowell "saw" were really optical illusions. What they actually saw were disconnected points on Mars that, when seen through Earth's turbulent atmosphere, they interpreted from a very human but not very scientific perspective.

The channels we actually see are often twisted and winding, and clearly show signs that liquid once flowed through them. The theory is that the liquid was water, probably released by cataclysmic events rather than in a continuous flow. For example, the channels near Hale crater in Mars's southern hemisphere (Figure 3) appear to come directly from the 90-mile (150-km) wide crater—evidence that water spewed out from the impact that created the crater. It could have come from melting ice below the surface. There is no similar evidence that water once flowed continuously over the Martian landscape, though the possibility is not discounted.

Figure 3: NASA's Spirit rover's wide-angle view on Mars. The large rock in the left foreground is Ulysses. To its right, there is a round feature that resulted when a drill on Spirit bored into the surface to make mineralogical measurements, which has been named Cyclops Eye. The Polyphemus Eye farther right does not show sulfate minerals, while Cyclops Eye does, leading to the conclusion that at the time of this photograph, about 2,000 days into the mission, Spirit was at a geological boundary.

Mars's atmosphere has been too thin for eons for liquid water to flow. But there might have been an era millions of years ago when Mars was warm and water flowed over its terrain. There is a perennial human hope that life arose during a warm Martian era and that some vestige of it, either alive or fossilized, might still be found. Most probably it would be underground, which is why Curiosity drilled into the Martian rock. We'll discuss the search for life on Mars later in this chapter.

From high-resolution images, we can see regions of Mars replete with layered rocks that look virtually identical to the sedentary rocks on our home planet. We know that the Earth rocks are formed under water. The NASA and ESA spacecraft are continually finding evidence that water once flowed on Mars.

The polar caps represent a major source of water on Mars. The caps are seasonal, but there is still a residual cap of water ice, even when the cap of carbon dioxide dissipates during that hemisphere's summer season. It is possible that there are vast quantities of water stored in a permafrost layer under the Martian surface.

The Martian Atmosphere

The orbiting spacecraft simultaneously map and study the thin Martian atmosphere. Mars's atmosphere is made up of 95% carbon dioxide—the same percentage as Venus's atmosphere. The key distinction is that Mars's atmosphere is only 1% the density at its surface as the Earth's atmosphere, and 1/10,000 the density of Venus's suffocating atmosphere. The Martian atmosphere is far too flimsy to support a greenhouse effect.

That is, Mars's atmosphere *now* could never support a greenhouse effect. Scientists have figured out that several billion years ago, the Martian atmosphere was much thicker. In that primordial age it might have had a greenhouse effect and a warm enough atmosphere for life to arise. If the loss of most of the atmosphere cooled the planet, then the situation on Mars would be the opposite of Venus's runaway greenhouse effect.

Comparative planetology provides us with interesting insights into the histories, present, and possible futures of different planets. Venus obviously shows us a possible future for Earth that we want to avoid. The weather patterns of Mars revealed by the orbiting spacecraft, the landers, and the Hubble Space Telescope are useful for helping us understand our weather patterns on Earth. For example, knowing how the global dust storms arise on Mars and spread dust over the planet can help us elucidate how dust from the Sahara Desert can sweep over the Atlantic Ocean to North America. After the volcanic ash from Iceland's Eyjafjallajökull ("island mountain glacier") spread across Europe in 2010, it drew worldwide attention to understanding the spread of volcanic dust.

The spacecraft on the Martian surface provide us with direct recordings of temperature. The temperatures range from a frigid -190 degrees F (-125 degrees C) at the Viking 1 lander's northern location to a balmy 80 degrees F (25 degree C) at the Viking 2 lander's site. The daily variation in temperature ranges from 60 to 90 degrees F (35 to 50 degrees C).

Scheduled for launch in later 2013, the main purpose of the Mars Atmosphere and Volatile EvolutioN (MAVEN) is to trace the history of the loss of Mars's atmosphere over time and so provide us with insight into how the Martian climate evolved. MAVEN is part of NASA's Mars Scout Program. Though the program was officially ended in 2010, MAVEN had been selected in 2008 as the 2013 Mars Scout Mission. MAVEN will reach Mars in 2014 to begin its measurements.

Roving Around

Beginning with Viking 1 and Viking 2, Mars landers have been exploring the ruddy terrain since 1976. The two Viking missions each had an orbiter and a lander. The landers stay fixed on the surface where they dig into the surrounding rocks and dust, transmitting meteorological observations back to Earth.

Subsequent missions to Mars carried landers that could move around. Tiny Sojourner looked like a child's toy, but was an object of tremendous fascination in those early days of the Internet. The hardy Spirit and Opportunity represent the next generation of Mars Exploration Rovers. They began their work examining the Martian surface in January 2004, for missions that were expected to last a few months each. But like the Energizer Bunny, they kept on going and going…

Opportunity is still going. Spirit became immobile in 2009 and stopped communicating in 2010, but Opportunity is still performing its work after nine Earth years. The initial assumption was that the solar panels, which provide the rovers with needed electricity, would be covered in dust over time, and the weakened electrical power would limit their lifetimes. Instead, the swirling dust from the sporadic dust storms and winds blowing across the Martian terrain actually *clean* the dust from the panels and revitalize the rovers.

Given Mars's distance of half an astronomical unit from Earth at its closest point, the radio signals commanding the rovers take at least four minutes to get there. If we had to rely on two-way communication to direct the rovers, it would take at least eight minutes and probably more at most times. Hence, the robots have self-contained capabilities for monitoring their own travel, which makes things much more efficient.

The rovers are equipped with tools for cutting into a rock or soil sample to examine its chemical composition and determine what minerals it contains. Rocks are cut with a rock abrasion tool, or RAT. In addition to the indigenous rocks, the rovers also turn up rocks that have landed on Mars from outer space.

Spirit and Opportunity have surpassed all expectations.

Figure 4: The Mars Exploration Rover named Opportunity uses its rock abrasion tool ("RAT") to cut a 2-inch circle into a rock named Marquette Island The Quest for Life on Mars.

From the Viking landers to Curiosity, all the Mars landers are equipped with instruments for detecting life in various forms. The Vikings were capable of taking in bits of Martian soil and wetting them or heating them to see what happens. What they were searching for were signs of respiration or metabolism. The general consensus was that the results were negative. There was one intriguing test result that seemed to signify life, but it was ultimately interpreted as a chemical reaction (no doubt to the disappointment of many).

Another experiment searched for organic material. Once again, the results were negative. Even if there had been some residual organic material on Mars, the Sun's ultraviolet might have destroyed it. Regardless, the findings were not encouraging for the idea that Mars once had life in the past.

When the European Space Agency launched its Mars Express mission in 2003, it included a last-minute, underfunded lander called *Beagle 2*, named for Darwin's ship, Beagle. After being released from its mother ship, Beagle was never heard from again. Mars Express is still orbiting, but Beagle probably crashed (its actual fate is unknown).

Phoenix, which landed on Mars in 2008, is part of NASA's Mars Scout Program, a series of relatively small, low-cost spacecraft. Landing on an arctic plain, Phoenix performed experiments for five months. The trenches it dug yielded water ice, along with chemicals and minerals in the soil demonstrating that the climate in that location was wetter and warmer within the last few million years—relatively recent for Mars. Phoenix found signs of perchlorate, a chemical that could mix with water to form brine that remains liquid at the temperature of the Martian surface. Perchlorate is compatible with some terrestrial microbes. It also made news as a rocket fuel component that has contaminated many water supplies on Earth. For astronauts, it has the potential to be used for rocket fuel or to generate oxygen for breathing.

The most ambitious and exciting mission to search for life on Mars is NASA's Mars Science Laboratory, launched on November 26, 2011. The spacecraft carried Curiosity, which successfully landed in Gale Crater on August 6, 2012. Too heavy to land the same way as its predecessors, Curiosity descended to the Martian surface powered by a novel sky crane system. The landing site has been named Bradbury Landing in honor of late science fiction author Ray Bradbury.

Curiosity immediately began its work in an area of Gale Crater called Yellowknife Bay, which looks like it once had a stream that flowed into a larger body of water. Rock dug up by Curiosity has been found to contain six elements that terrestrial microbes feed on. At some point in the distant past, they might have fed Martian microbes. Curiosity's work has just started and is sure to turn up many exciting discoveries.

Martian Moons

In his famous *Gulliver's Travels*, Jonathan Swift wrote that Mars had two moons. He couldn't have possibly known that in 1727, but it turned out that he was right. Mars has two small moons, named for Roman companions of the god of war. They are Phobos and Deimos, which translate into Fear and Terror, respectively. The moons were discovered in the nineteenth century with the U.S. Naval Observatory's telescope in Washington, D.C. They were no more than tiny points of light until their features were captured by close-ups from space.

Despite their formidable names, Phobos and Deimos are probably best described as big chunks of rock. Neither moon is even close to being round; they're both too small to have enough gravity to pull them into a spherical shape. Phobos and Deimos might be the sole survivors of an earlier satellite that was pulverized by a collision. Or, they might be captured asteroids.

Phobos is the larger of the two moons, but still diminutive, measuring 15 miles (27 kilometers) at its longest distance and 12 miles (19 kilometers) at its shortest. Phobos clearly bears the scars of those primordial space collisions. Its surface is covered with craters. One crater measures five miles (eight kilometers) across, taking up a sizable chunk of the small moon's diameter.

Tiny Deimos measures only nine miles (15 kilometers) at its longest point and seven miles (11 kilometers) at its shortest. Its appearance is much less craggy than Phobos. Dust seems to have partially filled in the craters, smoothing them out to our view. The dust covers Deimos's terrain and is roughly 30 feet (10 meters) deep.

Phobos and Deimos revolve around their Mars in periods of only two hours. Phobos completes its journey around the red planet in seven hours and 40 minutes, less than one-third of the Martian day of 25 Earth hours. In contrast, Deimos's orbit takes more than a Martian day, lasting 30 Earth hours. Because the speedier moon orbits faster than Mars rotates and the slower one lags behind it, if there were real Martians (or human visitors) standing on Mars's surface, they would see one moon moving forward overhead and the other moon moving backward.

Using the distance of Phobos and Deimos from the spacecraft and the Sun, and measuring how bright they appear from their images, we can calculate their albedoes, the percentage of sunlight they reflect. Each moon reflects only 7% or 8% of the sunlight hitting their surfaces; in other words, they are very dark. Their apparent brightness only comes from the contrast between the moons and the dark background. Their low albedoes suggest the moons are composed of rock rich in carbon. In that respect, they resemble some of the asteroids in the asteroid belt, which supports the theory that Phobos and Deimos might be asteroids captured by Mars.

Future Mars Exploration

Curiosity is hard at work, and MAVEN should tell us how Mars evolved from a warm and wet planet to a dust-covered desert. Missions to Mars are continually being proposed and planned by space agencies throughout the world. For human missions to Mars, we will have to wait at least another decade.

PART THREE

Gas Giants

 # Jupiter: King of the Solar System

In This Chapter

> Fast rotation and equatorial bulge

> The Great Red Spot

> Bands, belts, zones, and clouds

> Galileo orbiting Jupiter

> Jupiter's magnetic fields

Aptly named for the ancient king of the gods, Jupiter is the biggest, brightest, most massive planet in our solar system. With its orbit outside of Earth's, Jupiter can be high in the sky at any time of the day or night, depending on the point it occupies in its orbit. Only outshone by Venus as the brightest object in the night sky, apart from the Moon, when Venus sets or before it rises, Jupiter is the dazzling object that commands our attention.

Jupiter's dominance in the solar system is symbolized by the fact that the four gas giants—Jupiter, Saturn, Uranus, and Neptune—are sometimes called Jovian planets after the biggest one of the group. It also holds a prized place in the history of modern astronomy because Galileo's observations of Jupiter and its moons made a strong case for the heliocentric model of the solar system.

Jupiter is a favorite of amateur astronomers as some of its famous features can be seen with telescopes of all sizes. Galileo discovered that. Even small telescopes reveal bands across Jupiter's disk, as well as the changing positions of the four Galilean moons. Amateur observations of Jupiter are enhanced by cutting-edge techniques. Expert amateur astronomers snap hundred of pictures of Jupiter in quick succession. Computer programs select the ones with the clearest images and align them, producing wondrously detailed images.

Jupiter, photographed from the ground by a talented amateur astronomer, in an image that resulted from computer selection and stacking of many short images. We see many bands and zones across Jupiter, and we see its Great Red Spot. Other reddish spots, which are giant storms, also show. Since most telescopes invert images, the Great Red Spot, which is in Jupiter's southern hemisphere, appears in the upper half of the image. (credit: © Christopher Go; Cebu, Philippines)

Giant Jupiter has 318 times the mass of our small, rocky Earth, but that's still only one-thousandth the mass of the Sun. A planet like Jupiter would need roughly 15 times more mass to equal even the least massive star. Impressive as Jupiter is for our solar system, astronomers are discovering many planets as massive or several times more massive than Jupiter orbiting other stars.

Smart Facts

Jupiter Facts

Jupiter's orbit

Average distance from Sun	5.2 AU	778 million kilometers
Period by the stars	11.9 years	
Period from the Earth	399 days	
Orbit's eccentricity	0.05	
Orbit's tilt	1.3°	

Jupiter's features

Jupiter's diameter	142,984 km	011.2 times Earth's
Jupiter's mass	318 times Earth's	
Jupiter's density	1.33 times water	
Jupiter's surface gravity	2.54 Earth's	
Sidereal rotation period	9 hours 50 minutes to 9 hours 55 minutes	

Gas Giant versus Terrestrial Planets

In terms of size, Earth is a puny rock compared to the massive gas giant. Jupiter's diameter is 11 times bigger than Earth's. As volume depends on the cube of the diameter, Jupiter's volume is more than one thousand-seven hundred times Earth's. To put things in perspective, if our Earth was the size of a standard desktop globe about one foot across, Jupiter would be the size of a weather balloon measuring 11 feet across.

Smart Facts

Because it is outside of Earth's orbit, Jupiter can light up the night at any hour at different times. To the human eye, it shines two or three times brighter than Sirius, the brightest star. The four Galilean moons (the largest of Jupiter's 67 satellites) are visible through a telescope or binoculars. The faint bands of clouds on the Jovian surface are visible through even a small telescope.

Moving along the ecliptic, Jupiter is easy to find near the Hyades in Aldeberan in Taurus in 2013, Gemini in 2014, Cancer in 2015, and Virgo in 2016. Its oppositions, when it rises as the Sun sets and is highest at midnight, will occur next in January 2015 and about every 13 months after that.

According to a formula Newton figured out for Kepler's third law of planetary motion, we can calculate a planet's mass by following the period of a moon around its planet and measuring the size of that moon's orbit. (The assumption is that the satellite has negligible mass compared to the mass of the central body.) An object's density is its mass divided by its volume, for which the mass of water is 1. Jupiter's density is only 30% greater than that of water and only about 25% the density of Earth. From that measurement alone, it's obvious that Jupiter is not a terrestrial planet.

Moving out from the Sun, Mars is the last of the four terrestrial planets, the others being Mercury, Venus, and Earth. Jupiter is the first of the gas giants, the four outer planets. Its core is thought to consist of iron-rich material, about ten times more massive than the entire Earth. That core, however, is only a tiny fraction of Jupiter's mass.

To compensate for its low density, Jupiter is 85% hydrogen gas—the lightest of all the chemical elements. Almost all the remaining substance is helium, the second lightest element. All the other constituents of the giant planet, including methane and ammonia, account for less than 1% of Jupiter's mass. In composition, Jupiter is really more like the Sun and the other stars than terrestrial planets like Earth. Jupiter, though, is a lot cooler than any star, so we know that not all of its hydrogen is ionized or even in single-atom form. Most of the hydrogen in the Jovian outer atmosphere is made up of hydrogen molecules, H_2.

Descending from the tops of Jupiter's clouds, the gas keeps getting denser and denser until eventually it liquefies. There is no solid crust anywhere.

Jupiter and Ganymede · April 9, 2007
Hubble Space Telescope · WFPC2

NASA, ESA, and E. Karkoschka (University of Arizona)   STScI-PRC08-42

A view of Jupiter taken by the Hubble Space Telescope. Seen in this photograph are the Great Red Spot and some additional oval storms. Jupiter's Galilean moon Ganymede is about to move behind the planet.

Rapid Rotation

Jupiter is the fastest planet in the solar system in terms of rotation. The entire planet rotates in less than 11 hours, and some latitudes spin even faster. In fact, Jupiter spins so fast on its axis that it has a notable bulge at its equator. We call this shape "oblate," meaning that the diameter across its equator exceeds the diameter measured between its poles. Jupiter is oblate by about 7%. Its axis of rotation is only off 3 degrees with respect to its orbit, so there are no Jovian seasons.

The Hubble Space Telescope, telescopes on the ground, and space probes sent out from Jupiter all provide us with high-resolution images of the giant planet. The planet's speedy, uneven rotation spreads the clouds out into horizontal bands. Turbulence produces the details we see in the bands, which change from day to day.

A series of images taken with the Hubble Space Telescope show changes in the Great Red Spot over a 7-year interval.

The Great Red Spot

Whether seen from the Earth or from space, Jupiter's most prominent feature is the Great Red Spot. The "red spot" is actually a reddish-tinged hurricane that has been visible for more than three hundred years. It may have first appeared in 1665.

As devastating as hurricanes can be on Earth, once they reach land, they lose power and eventually die out. On Jupiter, there is no land to short-circuit their energy. Hurricanes like the Great Red Spot can go on indefinitely.

The reddish tinge comes from a small amount of a gas, probably a molecule containing phosphorus. We can also see several smaller colored spots on Jupiter, as well as a few white ovals. Videos captured by the Hubble Telescope and spacecraft show the revolution of the Great Red Spot. As it revolves it takes in bits of gas from adjacent bands.

The Great Red Spot is huge. It measures roughly 16,000 miles (26,000 kilometers) by 8,000 miles (about 13,000 kilometers) and changes size and shape from one year to the next. For perspective, the Earth's diameter is only about 8,000 miles. The storm extends upward for about five miles (eight kilometers), soaring over the tops of the surrounding clouds.

Located in Jupiter's southern hemisphere, the Great Red Spot has higher pressure than its surroundings and rotates in a counterclockwise direction. That makes it an anticyclonic storm. The spot seems to rotate roughly every six days.

Computer models show that when a planet has several rotating storms, they tend to merge. The biggest surviving storm keeps getting bigger. Jupiter's Great Red Spot embodies this effect. As different as Earth is from Jupiter, our planet has a few large-scale circulation patterns such as rings in the Gulf Stream. Studying the Great Red Spot can help us understand such phenomena in our terrestrial environment.

Bands and Clouds

Jupiter has bands of various types. The bright bands, called zones, are composed of falling gas. The darker bands, or belts, are made up of rising gas. The tops of the belts are roughly 12 miles (20 kilometers) lower and 20 degrees F (10 degrees C) warmer than the zones.

On Earth winds are fairly predictable. For example, the famous trade winds blow westward across low latitudes. At the middle latitude, a lesser known band of winds called the westerlies blows eastward. Jovian winds are a lot more complicated. On each hemisphere, a half dozen currents blow relatively eastward or westward with respect to Jupiter's (fast) average rotation.

As the various bands of wind blow over each other, small cells of gas start to circulate. On Jupiter, this effect generates storms and transforms small storms into more powerful ones.

Jupiter is replete with lightning storms, which are fascinating to look at. Many have been captured by space cameras, including a few superbolts. The lightning comes from regions where there are thunderheads. Notably, the symbol of the god Jupiter (Zeus in Greek mythology) was a thunderbolt.

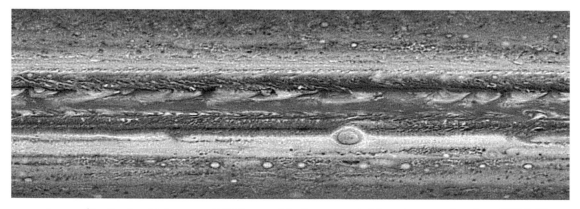

A map of Jupiter in which we see all 360° around Jupiter computed from images taken from NASA's New Horizons spacecraft when it flew by in 2007, en route to Pluto. Clearly shown are the Great Red Spot, many smaller spots, and a variety of cloud patterns.

Seen through telescopes on Earth or the Hubble Telescope in low Earth orbit, Jupiter appears basically upright. From those vantage points, it is impossible to get a good view of the poles. It is only from spacecraft that can look down on the poles that we get to see the polar circulation patterns. The horizontal band pattern from the lower latitudes does not extend to the poles. The bands break up into small eddies before they get there.

In 1995, NASA's Galileo spacecraft dropped a probe into the Jovian clouds. In view of Jupiter's enormous size, a result obtained from a single probe on a single path downward would not necessarily be typical. Indeed, the results did seem to be unusual. There was very minimal water vapor, which surprised the scientists. They surmised that the probe must have followed a trajectory that was abnormally dry.

That idea was supported by images taken by Galileo. The photographs showed towering clouds. The scientists theorize that the probe hit one of the hot spots that cover only about a scant 1% of Jupiter's surface. The downwelling air in these hot points is dry. It seems somewhat ironic that of the vast cloud formation, the probe would wind up in such an atypical spot, but that may very well be the case.

The probe lasted long enough to transmit data from about 400 miles (600 kilometers), or roughly 1% of Jupiter's radius. It confirmed that the percentage of helium was very similar to the Sun's, about 15%. The winds were stronger than the scientists anticipated, and they kept getting more powerful as the probe descended.

The tremendous pressure finally ended the life of the probe. It lasted until the pressure reached 24 times that of Earth's atmosphere, and then it died. A key conclusion drawn from the observations sent back is that the Jovian weather is supported by heating from below. On Earth, we have the opposite situation. Our weather is supported primarily by heat from the Sun, which comes from above.

Jupiter Missions

Terrestrial telescopic observations have always produced exciting views of Jupiter, but observations from space completely transformed our knowledge of the intricacies of the gas giant and its satellites. The first spacecraft to Jupiter were part of NASA's Pioneer series. These were the type of spacecraft that were stabilized by

Smart Facts

Missions to Jupiter

Pioneer 10, NASA, 1973
Pioneer 11, NASA, 1974
Voyager 1, NASA, flyby, 1979
Voyager 2, NASA, flyby, 1979
Galileo, NASA, orbiter, 1995-2003
Cassini, NASA, flyby en route to Saturn, 2000-2001 5-month encounter
New Horizons, NASA, flyby en route to Pluto, 2007
Juno, NASA, orbiter, launched 2011 to reach Jupiter 2016

rotation, so their pictures had to be recreated from data and were not the sharp, high-quality images produced by later generation spacecraft. But despite that limitation, they provided detailed images of Jupiter's disk and showed the unique characteristics of each of Jupiter's moons. Each of the many moons has its own personality! Seen from Earth, the satellites were no more than amorphous dots. Pioneer 10 flew by Jupiter in 1973 and Pioneer 11 followed the next year.

Later that decade, NASA launched Voyager 1 and Voyager 2, both with three-axis stabilization and therefore capable of taking high quality photographs. The Voyager pair flew by Jupiter in 1979, capturing stunning images as they took measurements of Jupiter's magnetic fields, tracked the influx of particles, and carried out numerous other measurement tasks.

The Voyager spacecraft made a tremendous contribution to our knowledge of the Galilean moons and discovered Jupiter's rings. We'll be discussing Jupiter's satellites and rings in the next chapters. We can also credit the Voyager missions with confirming the anticyclonic nature of the Great Red Spot.

Orbiting the Gas Giant

Entering Jupiter's orbit on December 7, 1995, Galileo is still the only spacecraft to orbit Jupiter. From its first orbit until its demise nearly eight years later, Galileo navigated various paths through the Jupiter system. Its different trajectories brought it close to each of Jupiter's many moons.

Galileo's first task on its arrival was dropping the probe into Jupiter's clouds (if not in the best spot). After that, it took close-ups of Jupiter and of the four Galilean moons, Io, Europa, Ganymede, and Callisto, and of Amalthea, the largest of the inner satellites (also called the Amalthea group).

Galileo was equipped with a suite of scientific instruments. It measured virtually every type of particle in its path and consistently took samples of the magnetic field. Taking accurate measurements of its position from Earth provided scientists with a much more precise picture of Jupiter's magnetic field, conducive to making more refined and accurate models of its interior. Monitoring Galileo's trajectories also enabled the scientists to analyze the gravitational effects on the spacecraft of Jupiter's myriad moons, which vary greatly in mass.

Tidal activity detected by the Voyager missions raised the tantalizing prospect that Europa, the smallest of the four Galilean moons (still huge by Earth standards), might have an ocean. Scientists were certain that there was liquid water under Europa's icy crust. Galileo's detailed measurements provided more evidence that Europa does have an ocean, making it a prime destination in the search for life that arose outside of Earth. Alas, that was to be Galileo's demise. Preventing any type of contamination of Europa was a top priority. To avoid even the remote possibility that Galileo could impact Europa, after its useful lifespan, the spacecraft was sent plummeting into Jupiter's atmosphere on September 21, 2003.

Jovian Magnetism

In the same way that the Van Allen belts of magnetic field surround the Earth, there are belts of magnetic field surrounding Jupiter. Of course, there is a vast difference in strength. Jupiter's magnetic field is 14 times as strong as our Earth's. Moreover, giant Jupiter has the most powerful magnetic field in our solar system (with the exception of sunspots).

Jupiter's inner magnetic field is shaped like a donut, like the Van Allen belts. Its outer magnetic field is enormous, and expands and contracts as it interacts with the solar wind flowing out from the Sun.

The huge magnetic field, generated by materials swirling deep in the gas giant's interior, endows Jupiter with auroras near its magnetic poles. The Hubble Telescope has captured stunning images clearly displaying the auroral ovals, which grow larger and stronger when fed by particles from the solar wind.

Juno

Juno, the Roman queen of the gods, is the name of NASA's latest venture to Jupiter. In Greco-Roman mythology, Jupiter cloaked himself in a veil of clouds to hide his wayward behavior. Juno (Hera in Greek), his wife, was the only one who could see through the clouds to reveal Jupiter's true nature.

Juno was launched on August 5, 2011. In 2016, the New Frontiers spacecraft is expected to go into polar orbit around the planet, studying its atmosphere, composition, gravitational field, magnetic field, and magnetosphere. Infrared and radio wave measurements of radiation deep inside Jupiter's atmosphere will help determine how much water is in the clouds. Water implies that there is oxygen present. In studying Jupiter's interior, Juno will explore whether the gas giant might have a rocky core.

Studying Jupiter goes way beyond giving us information about the planet itself. Scott Bolton, Juno's lead investigator from the Southwest Research Institute in San Antonio calls Jupiter "the Rosetta Stone of our solar system." Much, much older than any other planet and infused with "more material than all the other planets, asteroids and comets combined," Jupiter can unlock the mysteries of the origin of our solar system.

Juno has one limitation. For stabilization, it will be spinning two to five times per minute, which makes it more like the Pioneers than the recent generations of spacecraft. But if it will not be providing us with stellar visible-light images, it will be sending back a wealth of knowledge of the planet that dominates our solar system and of the solar system itself.

CHAPTER 10

Jupiter's Galilean Moons: Like a Mini-Solar System

In This Chapter

➤ Galileo sees stars near Jupiter, realizes they are four moons

➤ Io, young and volcanically active

➤ Europa, home of an ocean under the ice

➤ Ganymede, the biggest of all the moons

➤ Callisto, covered in craters

During the year 1609, the ever-inventive Galileo was busy improving the observational quality of his telescope. With superior magnifying ability, the improved version enabled him to hone in on distant celestial objects with greater sharpness and clarity. Watching the heavens from Padua, Galileo saw bright points of light near Jupiter. He initially thought they were stars. But these "stars" followed a particular pattern: their positions changed in relation to Jupiter but they always stayed in close proximity to the huge planet. Soon Galileo realized he was looking at moons belonging to Jupiter.

Galileo first mentioned the bright objects in a letter written January 7, 1610. At that time, he saw only three presumed stars. As his observations continued from January 8th through March 2nd, he discovered a fourth object and recognized that the four bodies were orbiting Jupiter, like our own Moon orbits the Earth. And even more significant than the discovery per se, the fact that these objects were *not* orbiting Earth, but rather another celestial body, was a massive blow to the geocentric theory and a major boost to Copernicus's heliocentric view of the universe.

Solar System Scoop

Not quite as revolutionary, but no less important for modern astronomy, Galileo's discovery of Jupiter's satellites also confirmed the value of the telescope as a tool for revealing objects in the sky that could not be seen with the naked eye. Four hundred years after Galileo turned his lenses skyward, a commemorative plaque was placed atop the Campanile, a 15th-century tower (recreated after it collapsed in 1902) in Venice. The commemoration was actually for Galileo's showing the nobles of Venice how his invention could be used to see ships fairly far out at sea. However, the amazing discoveries it made possible are implicit in the inscription. Approximately translated, the plaque reads: "Galileo Galilei, with his spyglass, on August 21, 1609, enlarged the horizons of man. 400 years ago."

Naming the Moons

In Renaissance Venice, patronage was the name of the game. Galileo had been the mathematics tutor of Cosimo de' Medici, who was awarded the title of Grand Duke Cosimo II of Tuscany in 1609. The discovery of Jupiter's moons seemed a marvelous opportunity to gain even more favor with the wealthy scion of a very powerful family, now Galileo's patron. In a letter to Cosimo's secretary, written February 13, 1610, Galileo proposed immortalizing the Grand Duke by naming his discovery the "Cosmian Stars," after Cosimo, or the "Medician Stars," honoring all four Medici brothers.

Cosimo himself preferred the Medician Stars (*Medicea Sidera*). And so it was…briefly. *Sidereus Nuncius* was published in March 1610, with a dedication exalting Galileo's renowned patron and elaborating the "four stars reserved for your illustrious name." The Medici family is indeed immortalized in history, but not for Jupiter's moons.

There were several rival names proposed for the satellites. Ultimately, Simon Marius won out with the names we know the moons by today. (Marius might even have seen the moons before Galile, but his writing appeared much later.) Ancient mythology prevailed once again. The four moons were named Io, Europa, Ganymede, and Callisto—all seduced by the god Zeus, the Greek counterpart of the Roman Jupiter.

To commemorate the 400th anniversary in 2009 of Galileo's first use of the telescope on the heavens, a plaque was put on a column at the top of the Campanile, a 15th-century tower in Venice. The commemoration is of Galileo's demonstrating his telescope to the nobles of Venice by observing ships far out at sea; it was before he turned his telescope upward. The writing on the plaque can be translated as "Galileo Galilei, with his spyglass, on August 21, 1609, enlarged the horizons of man, 400 years ago."

Solar System Scoop

Zeus was notorious for his amorous activities. *Io* was a priestess of Zeus's wife Hera, whom Zeus seduced and then transformed into a heifer to hide her from Hera. Hera was not fooled. She sent a gadfly to sting the heifer Io, who eventually wandered to Egypt.

Europa was seduced by Zeus, who disguised himself as a bull and carried her off to Crete. Homer described this scene in *The Iliad*. Europa is said to be a descendent of Io.

Ganymede was an extremely handsome young Trojan prince. This time, Zeus took the form of an eagle, swooping down on Ganymede to abduct him and bring him to Mount Olympus. There, Ganymede served as cupbearer to the gods and was one of Zeus's lovers.

Callisto, daughter of the king of Arcadia, had taken a vow to remain a virgin. As Ovid tells it, Zeus disguised himself as the goddess Artemis to seduce her. After Callisto gave birth to a son, Arkas, a jealous goddess, turned her into a bear. Arkas killed the bear with an arrow. Mother and son were both immortalized in the sky as the Great Bear and the Little Bear, Ursa Major and Ursa Minor.

Galileo was *not* happy with Marius's names for the moons, and he adamantly refused to use them. Instead, he created a numbering system that is still used today in conjunction with the moons' names. The numbers are based on the moons' proximity to Jupiter, getting higher as the moons extend outward. So Io, Europa, Ganymede, and Callisto are I, II, III, and IV, respectively.

Galileo would probably not be surprised that the numbers of Jupiter's moons keep increasing as powerful ground-based and space-based telescopes make new discoveries of small, faint satellites orbiting Jupiter. (We'll discuss those smaller moons, along with Jupiter's rings, in the next chapter.) But he might be amazed to see the magnificent, detailed images of the "Medician Stars" brought back to Earth by the spacecraft that bears his name and its predecessors.

Galileo himself would have been amazed to see what his namesake spacecraft and its predecessors showed from the "Medicean satellites" that he discovered in 1609. From left to right these photographs show Io, newly resurfaced with two dozen continually erupting volcanoes. Second is Europa, a prime suspect in the search for extraterrestrial life because of the ocean that is under the smooth ice layer that we see. Third is Ganymede, the largest moon in the solar system (omitting atmosphere in the calculation), showing especially a fascinatingly grooved part of its surface. And at right is Callisto, farther out than the others and covered with hard ice that retains the scarring from overlapping meteorite strikes that transpired over billions of years.

Smart Facts

Orbits of the Galilean Moons

		Orbital radius (km)	Eccentricity	Tilt	Period (days)
Io	1822 km	421,800	0.0041	0.036°	1.769
Europa	1561 km	671,100	0.0094	0.466°	3.551
Ganymede	2631 km	1,070,400	0.0013	0.177°	7.155
Callisto	2410 km	1,882,700	0.0074	0.192°	16.69

Io

Io might be more accurately named "Vulcan." The innermost moon, the reddish tinged Io is the fourth largest moon in our solar system—and the most geologically active object of any type. When Voyager flew by, engineer Linda Morabito of the Jet Propulsion Laboratory in Pasadena, California, scrutinized a very long exposure of Io, shot to bring out background stars for navigational purposes. Something appeared to be protruding off Io's disk. It was a huge volcano.

Io is (literally) a hotbed of volcanic activity. Its landscape is covered with mountains and more than 100 are active volcanoes. Its biggest volcano, Tvashtar, spews lava upward 190 miles (300 kilometers).

Detailed images of Io show strange-looking splotches that are actually fields of lava. Voyager 1 captured images of eight erupting volcanoes. Many were still erupting several months later when Voyager 2 flew by. Galileo tracked Io's hyperactive volcanic activity as it orbited the Jupiter system from 1995 to 2003. Since then, monitoring Io's volcanoes has been taken over by the Hubble Space Telescope and any spacecraft that passes by. The most recent one was New Horizons en route to Pluto.

Io, the innermost of the four Galilean satellites has an interior that is very active, from tidal heating, leading to dozens of erupting volcanoes that resurface its crust often. Io's surface is the youngest of a solar-system object. In the New Horizon spacecraft's passage by Io, shown here at top with two volcanoes erupting hundreds of miles off its edge, it found changes on the surface since Galileo had photographed it in 1999. The biggest volcano, Tvashtar, rises 190 miles. Prometheus is at left and Amirani is between them.

Before the Pioneer missions to Io, theoretical calculations pointed to Io having a very hot interior. Tidal forces heat the interior, producing the plethora of volcanoes forever spewing lava and continually resurfacing the moon's crust. As a result of this ongoing lava flow, Io has the youngest surface of any object in our solar system.

Additional forces make Io's interior even more of a furnace. Gravitational tugs from Jupiter's other moons, as well as the powerful planet itself, flex Io's surface, causing it to go in and out by roughly 300 feet (100 meters)—the size of a football field. This continual flexing further heats up the interior. Though scientists were aware of this phenomenon, they were still amazed at the scope of volcanic activity, as the images clearly revealed volcanoes shooting lava for hundreds of miles. Io was known to be hot, but it was not expected to be volcanic, and certainly not to such a tremendous extent.

The discovery of Io's volcanoes solved another intriguing mystery. Scientists knew that Io was giving off particles that went into orbit around Jupiter. These unusual particles formed a doughnut shape (*torus*) at Io's orbital radius, known as the *Io torus*. What was not initially known was the source of these particles—which proved to be the volcanoes. Potassium, sulfur, sodium, and oxygen have all been found in the Io torus.

Far out in space and dotted with raging volcanoes, Io is the antithesis of a Goldilocks planet. Io's average surface temperature is about -240 degrees F (-150 degrees C). The volcanic hot spots can soar to 1,800 degrees F (1,000 C).

And those hot spots may be cool compared to what Io must be like underground. Io is thought to have an underground ocean of liquid sulfur (surely an image from Dante's *Inferno*). When sulfur cools, it assumes various colors, which would explain the weirdly colorful appearance of Io's landscape.

Io has mountains that dwarf Mount Everest, with peaks rising as high as 10 miles (16.1 kilometers). Since sulfur mountains would collapse at that height, they are probably built on silicon and coated with sulfur. Yet another one of Io's idiosyncrasies—most of the outer objects in our solar system have surface features coated with ice.

Based on its gravitational pull on the Galileo spacecraft, Io's density was shown to be 3.5, more than three and one half times the density of water. Consequently, its core must be composed of heavy elements. Io is presumed to be rich in iron and perhaps iron sulfide. Its core could be either one: iron sulfide or molten iron.

Appropriately, the names of Io's features are all mythological characters connected in some way with fire. Two of the most notable are volcanoes named for the Greek Prometheus, who brought fire to humankind, and Pele, the Hawaiian volcano goddess.

Europa

Io and Europa are like two opposite sisters. Or specifically, like fire and ice. Icy Europa's smooth, glossy landscape contrasts sharply with hot, active, mottled Europa. The icy nature of Europa's surface is shown by its high reflectivity (albedo). Infrared spectra confirmed that the surface is ice. Cracks in Europa's surface resemble the cracks in blocks of ice in the polar oceans on Earth.

Europa, the second innermost of the four Galilean satellites, with its very smooth, icy surface is shown here in a composite photograph.

Europa is the second innermost of the four Galilean moons and the smallest, slightly smaller in size than our own Moon. With very few craters, it is one of the smoothest objects in our solar system. The ice is probably soft enough under the crust for the craters to disappear under the surface, which is tectonically active and young. Europa's interior might have been subject to less internal heating than Io's, or alternately, internal radioactivity might have contributed to the energy of Europa's interior.

High-resolution photographs taken by Galileo show what are apparently blocks of ice. This immediately raised two key questions: How thick is the ice? Could it be so thick that there is minimal or no interchange between the atmosphere of Europa and what lies below it?

The predominant opinion is that the ice really is very thick and almost impenetrable. Still, there are also dissenters who claim that the ice is thin enough to be permeable. If the dissenters are right, that would make it easier for a spacecraft to drill through the surface or use other techniques to sample under the ice. At the present time, though, the question is moot for practical purposes. The next planned mission to Jupiter's moons is not scheduled for launch until 2022.

An issue on which there is general agreement is that Europa has an ocean. There was long thought to be a water ocean under Europa's frozen surface. In a very exciting discovery, that idea was confirmed by Galileo's measurements of the magnetic field. The device detected changes in the magnetic field matching changes we would expect from an underground ocean, a liquid presence in the interior that conducts electricity. The main—and virtually the only—implication is that under all that ice is a salty, liquid ocean.

In the quest to find life beyond Earth, any sign of an ocean triggers excitement. Europa's apparent ocean has given it high priority as a destination for future exploration. As enticing as that prospect might be, we will have to wait another decade for the next Jovian mission.

Solar System Scoop

The lure of a possible liquid ocean has made not only Europa, but also Ganymede and Callisto prime targets for detailed exploration. Unfortunately, funding issues have disrupted plans to examine the Jovian moons. NASA's Jupiter Icy Moons Orbiter (JIMO) was cancelled in 2005. That was followed by plans for a joint NASA/ESA venture, Europa Jupiter System Mission-Laplace (EJSM-Laplace), which would have combined NASA's Jupiter Europa Orbiter with ESA's Jupiter Ganymede Orbiter. That partnership ended in 2011.

Enter JUICE. The **JU**piter **IC**y Moon **E**xplorer began as a revamped version of the Jupiter Ganymede Orbiter plan. ESA's planned Jovian mission will examine Ganymede, Europa, and Callisto in detail on the theory that the three Galilean moons all have liquid oceans under the surface, meaning they can provide us with invaluable knowledge of the potential of icy worlds to support life. As the title of the original project implies, Ganymede is the main focus of exploration, but the full mission will encompass all three potentially habitable moons.

Equipped with a suite of instruments, JUICE is scheduled for launch in 2022 and expected to reach the Jupiter system in 2030. After performing activities around Jupiter, Callisto, and Europa, JUICE should go into orbit around Ganymede in 2033. For now, we have Juno en route to reach Jupiter in 2016.

Ganymede

Ganymede holds the status of being the largest moon in our solar system. Its radius of 2,631 kilometers makes Ganymede bigger than Mercury (2,440 kilometers), not to mention our own Moon (1,738 kilometers). As an icy orb, however, Ganymede has only about half of Mercury's mass.

Ganymede's surface varies from place to place. Some regions are dotted with craters, while others are covered with ridges. On the side of Ganymede facing backwards as it orbits Jupiter (Figure 5), there are visible polar caps, along with the two main types of terrain. One type is bright and grooved, and the other is darker with notable furrows.

Ganymede is the largest satellite in the solar system, larger than Mercury or our Moon. Some regions are covered with craters and others with ridges. In this photograph we see the side that faces backward as it orbits Jupiter with the polar caps as well as the two major types of terrain on Ganymede. One type is bright and grooved. The other, darker, has furrowed areas.

Smart Facts

It is probably not surprising that "Galileo" is a very popular name in modern astronomy. *Galileo Regio* is the name of the largest dark region on Ganymede. The title exemplifies the tradition of naming similar dark features of natural satellites after astronomers.

In contrast to Io's young surface, Ganymede's surface is very old. The ice is so hard that like rock on terrestrial planets, it has impact craters covering roughly one-third of its frigid terrain. The cratered part of the landscape is darkened by dust that has been falling on Ganymede for billions of years. In some areas, we can see a bright crater and even some rays, a result of a fairly recent collision that revealed water ice under the dust. Ganymede is composed primarily of silicate rock and water ice. To highlight how icy it is, Ganymede's frost can extend from the poles to latitudes as low as 25 degrees. On Earth, that would be like a polar frost extending to Florida.

Thanks to Galileo's mapping of altitudes on Ganymede, we know that the youngest and smoothest regions can drop about 600 feet (200 meters) in elevation. A plausible explanation is that ice melted inside due to tidal forces and some of the water flooded these lower areas.

Ganymede has some unique features for a natural satellite. Its surface shows features that have shifted sideways, analogous to the continental plates on our home planet. In other words, Jupiter's biggest moon has its own version of California's San Andreas fault. Such faults only exist on Earth and on Ganymede. JUICE's detailed study of Ganymede may aid our understanding of terrestrial earthquakes.

Ganymede is the only natural satellite with a magnetosphere. Its magnetic field was discovered by an instrument aboard Galileo. This discovery implies that there is some internal circulation inside the Jovian moon. And of course, there is the prospect that Ganymede may have an ocean. If there is an ocean beneath the ice, it would be at least 100 miles (170 kilometers) underground. We may have to wait for JUICE to reach Ganymede to find out.

Callisto

Callisto is the least remarkable of the four sibling satellites. The outermost Galilean moon, Callisto lacks Io's volatility and Europa's smooth, icy gloss, and is not the biggest moon in the solar system like Ganymede. Size-wise, though, Callisto is only 10% smaller than Ganymede and is the solar system's third largest moon.

Callisto's hallmark is craters. Its frozen, hard surface is so heavily covered with craters dating back billions of years, that it's said to be *saturated* with craters. That is, there is no place on all of Callisto where another crater would alter the overall distribution.

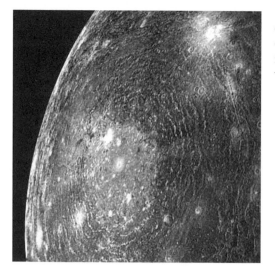

Callisto, the outermost of the four Galilean satellites, has an iron-hard icy surface that is entirely covered with craters.

Solar System Scoop

The most massive impacts formed the immense feature known as Valhalla, named for the majestic hall of the dead in Norse mythology. Cold, distant Callisto with an ancient dead surface seems an ideal place for Valhalla. Valhalla was ruled by the chief god Odin, Jupiter's Norse counterpart.

Discovered by Voyager 1, Valhalla consists of a huge set of at least 10 rings. A tremendous impact blasted the center, thawing the icy surface enough so that the waves radiated out over it before freezing in place. Material from the original crash is still visible in the inner rings.

Valhalla's rings are not entirely concentric, which distinguishes them from ringed features on Mercury and our Moon. Callisto also has a few smaller sets of rings on its surface.

Scientists examining photographs sent by Galileo were surprised to discover that not all Callisto's craters were visible. In particular, they expected to see more small craters. It is possible that the smallest craters (that is, smaller than football fields) might have been buried in dust from later collisions. Alternately, they might have self-destructed from static electricity.

Galileo's tracking of Callisto's gravitational field led to another unexpected discovery. The satellite's mass turned out to be more centrally concentrated than the scientists thought. The concentration implies that the heaviest materials have sunk. And what caused that could be a liquid ocean deep in Callisto's interior. Studies of Jupiter's magnetic field near Callisto suggest the presence of an interior ocean. JUICE's voyage is based on theory that Ganymede, Europa, and Callisto are all harboring liquid oceans under their ice, and thus the potential for some form of life.

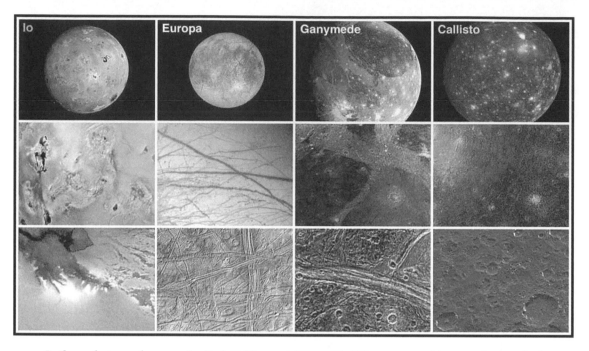

In these photographs we see Jupiter's Galilean satellites with different zooms. There are features in the top row as small as 20 kilometers in size. In the middle row, the resolution is up to 10 times higher. On Io, there are the black volcanic calderas. On Europe, there are cracks in the surface ice. On Ganymede, there are strangely grooved regions crossing smoother regions. On Callisto, there are huge impact basins that resulted from collisions with comets or asteroids. In the bottom row, there are examples of the highest available resolution images from Galileo on these objects. On Io, there are individual volcanic vents. On Europa, there are ridges in the ice. On Ganymede, there are what is called the grooved terrain.

CHAPTER 11

 # Jupiter's Numerous Rings and Moons

In This Chapter

> Giant planet, flimsy rings

> Myriad moons

When Galileo sharpened the view of his telescope lens, he successfully zeroed in on Jupiter's four biggest moons. He had less success in discovering rings, even on Saturn. Though he recognized something odd in his hazy image of Saturn, he had no idea he was looking at rings. Saturn appeared to have "ears." Jupiter had the four "Medician Stars."

It was not only Galileo's primitive telescope that failed to detect Jupiter's rings. By the mid-seventeenth century, Saturn was known to have rings. Even the rings around Uranus were discovered by terrestrial telescope. That discovery, by astronomers at the Kuiper Airborne Observatory in 1977, and combined with the knowledge of Saturn's rings, convinced astronomers that giant Jupiter would probably have a ring or rings of its own. Consequently, the Voyagers were programmed to search the Jovian system for rings. Their quest was fully rewarded. The Galileo orbiter and the Hubble Space Telescope have brought us detailed images of the Jovian rings, which are so faint they can only be seen by the most powerful telescopes on Earth.

Nearly 370 years passed between Galileo's observations of Jupiter's four largest satellites and the discovery of the rings in 1979. It took 282 years before the fifth-largest satellite, Amalthea, was discovered at the Lick Observatory in California in 1892. Jupiter's entourage now includes 67 moons.

The Jovian Ring System

The Jovian rings were first detected by Voyager 1 as it flew by the Jupiter system. With longer exposures, Voyager 2 built on and enhanced the discovery. Jupiter may dominate the other planets in size, but definitely not in rings. Unlike Saturn's magnificent rings, Jupiter's rings are skinny and faint. Composed mainly of dust, the rings are constrained in size by the gravity of the small moons orbiting just inside and outside the rings. The moons' gravitational pull keeps the ring material firmly in place.

One of Galileo's two surviving telescopes.

The ring system has four key components. The "halo ring," a thick inner torus, glows bluish in visible and near-infrared light, in contrast to the other rings, which have a reddish tinge. Next, there is the very skinny "main ring" which is fairly bright. The halo and main rings are made up of dust thrown from the moons Metis and Adrastea, and some tiny, unobservable bodies during those high-impact crashes in the early solar system. The two outer "gossamer rings" are composed of material ejected by Amalthea and Thebe, and are named for those moons.

The images provided by Voyager 2 provided a rough draft of the Jovian ring structure. Galileo's superior high-resolution images sharpened our knowledge tremendously, and photographs from the Hubble Space Telescope and the ultra-powerful ground-based Keck telescope illuminated the intricacies of the ring structure as seen in back-scattered light (that is, light reflecting back from particles in the field of view of the lens). In forward-scattered sunlight, which enhances the brightness of dust, the Jovian rings all shine brightly.

In 2007, as New Horizons flew by the Jupiter system, it sent back the first detailed image of the delicate structure within the main ring. As we discussed in the previous chapter, we have a long wait for the next Jovian mission.

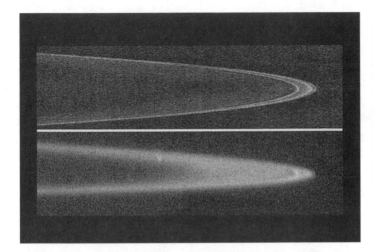

Jupiter's rings as imaged from NASA's New Horizons spacecraft as it flew by in 2007. The top image was taken during the spacecraft's approach, with back-lighted material whose size ranged from gravel to boulders. The lower image was taken as New Horizons departed, which shows the view of forward-scattered sunlight. which enhances the brightness of dust.

Moons and More Moons

Befitting its status as King of the Solar System, Jupiter has the largest retinue of confirmed moons. Unlike the long wait for Amalthea's discovery, the twentieth century produced a succession of Jovian moons. The discovery of Himalia in 1904 was followed by Elara in 1905, Pasiphae in 1908, Sinope in 1914, Lysithea and Carme in 1938, Ananke in 1951, and Leda in 1974. These discoveries by terrestrial telescopes were made by astronomers at the Mount Wilson, Mount Hamilton, and Palomar Observatories in California, with one from the Greenwich Observatory in England.

Also preceding Voyager, Themisto was initially sighted in 1975, but inadequate observation data rendered it lost until 2000. Thanks to Voyager, Metis, Adrastea, and Thebe joined the entourage in 1979.

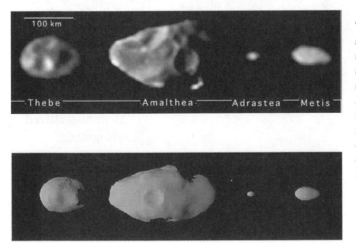

These four innermost satellites of Jupiter are all much smaller than the Galilean satellites. They are shown here in their actual relative sizes to each other. Metis, at left, is only 37 miles across. Tiny Adrastea is only 12 miles across. Amalthea is 154 miles across. Thebe, at right, is 72 miles across. The bottom set of images are computer simulations of the actual shapes of the satellites, based on a variety of spacecraft photographs.

Two decades went by before any new moons were found. Then, between 1999 and 2003, the discoveries were fast and furious. Thirty-two new moons were discovered—and by telescopes based on Earth.

Most of the new discoveries came from the team led by Scott S. Sheppard and David C. Hewitt of the University of Hawaii. The Canada-France-Hawaii telescope at the summit of Mauna Kea allows astronomers to see moons as tiny as one mile in diameter. And these moons are indeed *small*. The biggest one measures 5.6 miles (9 kilometers) across. The average diameter is 1.9 miles (3 kilometers). Sheppard and Hewitt are credited with discovering 43 of the current tally of 67 Jovian moons.

Compared to the miniscule new discoveries, Amalthea, at 154 miles (237 kilometers) across is a giant. At 53 miles (85 kilometers) in radius, Adrastea is roughly a scant 3% of the radius of Ganymede or Callisto, but still large by the standards of Jupiter's (non-Galilean) moons. Thebe is 72 miles (116 kilometers) across, and Metis is 37 miles (60 kilometers) across.

For perspective, "big" Amalthea is about the size of Long Island, New York. Ganymede and Callisto surpass our own Moon in size and are not all that small in comparison to our planet.

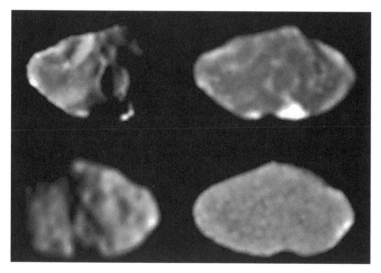

Amalthea is shown from NASA's Galileo mission. One face of Amalthea always faces Jupiter, because of the way the planet's gravity pulls on its irregularly shaped mass. Here the side facing Jupiter is shown in the top pair of images. This "leading side" shows a very large crater with respect to the size of the salellite itself. The brightest points, near the bottom of the image, are in another huge crater. The topography varies less in the "trailing side," shown in the bottom pair of images.

Solar System Scoop

Zeus's (or Jupiter's) amorous adventures leave no shortage of names for naming the Jovian moons after the god's favorites, lovers, and daughters. The use of various names in mythology for the same figure also broadens the pool. But it was not until the 1970s that the smaller moons got their names. Until then, only Amalthea, the first discovery of the innermost moons, had a name (although unofficially). *Amalthea* was the foster mother of Zeus, sometimes portrayed as a goat who suckled the infant Zeus.

In 1975, the International Astronomical Union's Task Group for Outer Solar System Nomenclature officially granted names to the non-Galilean moons and created a formal structure for naming future discoveries. There are several stories associated with some of the mythological characters. For example, in one legend, *Thebe* was the daughter of a river god loved by Zeus, and in another, the daughter of Zeus and Iodame. *Adrastea* was also the daughter of Zeus, according to one myth. Alternately, she was the daughter of the king of Helicon, and she and her sister Ida were Zeus's baby nurses, entrusted with keeping him safe from his murderous father. *Metis* predates Hera as an early and powerful wife of Zeus. Adrastea, Thebe, Metis, and Amalthea are known as the inner satellites or the "Amalthea group."

The more recently the moons have been discovered, the more esoteric their names seem to be. Those moons discovered since 2003 have yet to be named and are known only by "S" for satellite with the year of discovery, "J" for Jupiter, and a number (for example, *S/2003 J 2*).

The Galilean moons and the inner satellites are the "regular satellites" of the Jovian system, with prograde and almost circular orbits. The "irregular satellites" are the army of tiny moons with more distant and eccentric orbits. These are classified according to prograde or retrograde revolution and most are grouped into "families that share certain characteristics" (such as inclination, eccentricity, and semi-major axis). For example, the *Himalia group* is a prograde family, while the *Carme, Ananke,* and *Pasiphae groups* are all retrograde families.

The small Jovian moons are survivors of tremendous impacts from small asteroids and comets. Their assault was intensified by Jupiter's powerful gravity, which accelerated the speed and force of the battering objects. Given this history, it is not surprising that the retrograde families outnumber the prograde families.

None of the non-Galilean moons is massive enough for gravity to have pulled them into the round (or roundish) shape we associate with the word "moon." Despite this, some of the moons may have internal activity. For example, some white and highly reflective material has been observed near the crater Gaea (Mother Earth) on Amalthea.

Ironically, the best images of the smaller moons were taken by Galileo on its swan song, when it plunged into Jupiter. The images reveal an extremely varied collection of rings and satellites…and the potential for more of them yet undiscovered.

Characteristics of the Non-Galilean Moons
Orbital parameters

Satellite	Orbital Radius (km)	Eccentricity	Inclination (deg)	Period (days)
Himalia	11461000	0.1623	27.496	250.56
Elara	11741000	0.2174	26.627	259.64
Pasiphae	23624000	0.4090	151.431	743.63
Sinope	23939000	0.2495	158.109	758.90
Lysithea	11717000	0.1124	28.302	259.20
Carme	23404000	0.2533	164.907	734.17
Ananke	21276000	0.2435	148.889	629.77
Leda	11165000	0.1636	27.457	240.90

Satellite	Orbital Radius (km)	Eccentricity	Period (days)
Callirrhoe	24102000	0.2829	758.77
Themisto	7507000	0.2424	130.02
Magaclite	23808000	0.4211	752.86
Taygete	23363000	0.2518	732.41
Chaldene	23179000	0.2514	723.72
Harpalyke	21104000	0.2261	623.32
Kalyke	23564000	0.2464	742.06
Iocaste	21272000	0.2151	631.60
Erinome	23283000	0.2656	728.46
Isonoe	23231000	0.2465	726.23
Praxidike	21148000	0.2302	625.39
Autonoe	24033000	0.3166	761.00
Thyone	21192000	0.2377	627.19
Hermippe	21300000	0.2123	633.90
Aitne	23315000	0.2655	730.10
Eurydome	23148000	0.2758	717.34
Euanthe	21038000	0.2308	620.44

Satellite	Orbital Radius *(km)*	Eccentricity	Period *(days)*
Euporie	19339000	0.1440	550.69
Orthosie	21164000	0.2780	622.55
Sponde	23790000	0.3132	748.31
Kale	23302000	0.2521	729.55
Pasithee	23090000	0.2668	719.46
Hegemone	23566000	0.3439	739.85
Mneme	21036000	0.2271	620.06
Aoede	23969000	0.4324	761.41
Thelxinoe	21165000	0.2194	628.10
Arche	23355000	0.2555	731.90
Kallichore	23273000	0.2424	728.28
Helike	21064000	0.1467	626.32
Carpo	17078000	0.4436	456.25
Eukelade	23322000	0.2670	730.32
Cyllene	23787000	0.4175	752.01
Kore	24486000	0.3315	776.60
Herse	23405000	0.2493	734.55
S/2003J2	28332000	0.4114	979.33
S/2003J3	20230000	0.2027	583.87
S/2003J4	23928000	0.3558	755.22
S/2003J5	23493000	0.2459	738.75
S/2003J9	23385000	0.2641	733.31
S/2003J10	23042000	0.4284	716.28
S/2003J12	17835000	0.4884	489.66
S/2003J15	22622000	0.1868	689.77
S/2003J16	20948000	0.2309	616.33
S/2003J18	20494000	0.1016	598.11
S/2003J19	23532000	0.2620	740.39
S/2003J23	23549000	0.2702	732.47

Physical parameters

Satellite	Mean radius *(km)*	Mean density *(g/cm3)*	Geometric Albedo
Io	1821.6 ± 0.5	3.528 ± 0.006	0.63 ± 0.02
Europa	1560.8 ± 0.5	3.013 ± 0.005	0.67 ± 0.03
Ganymede	2631.2 ± 1.7	1.942 ± 0.005	0.43 ± 0.02
Callisto	2410.3 ± 1.5	1.834 ± 0.004	0.17 ± 0.02
Amalthea	83.45 ± 2.4	0.849 ± 0.199	0.090 ± 0.005
Himalia	85	2.6	0.04
Elara	43	2.6	0.04
Pasiphae	30	2.6	0.04
Sinope	19	2.6	0.04
Lysithea	18	2.6	0.04
Carme	23	2.6	0.04
Ananke	14	2.6	0.04
Leda	10	2.6	0.04
Thebe	49.3 ± 2.0	3.0	0.047 ± 0.003
Adrastea	8.2 ± 2.0	3.0	0.1 ± 0.045
Metis	21.5 ± 2.0	3.0	0.061 ± 0.003
Callirrhoe	4.3	2.6	0.04
Themisto	4.0	2.6	0.04
Megaclite	2.7	2.6	0.04
Taygete	2.5	2.6	0.04
Chaldene	1.9	2.6	0.04
Harpalyke	2.2	2.6	0.04
Kalyke	2.6	2.6	0.04
Iocaste	2.6	2.6	0.04
Erinome	1.6	2.6	0.04
Isonoe	1.9	2.6	0.04
Praxidike	3.4	2.6	0.04
Autonoe	2.0	2.6	0.04
Thyone	2.0	2.6	0.04
Hermippe	2.0	2.6	0.04
Aitne	1.5	2.6	0.04
Eurydome	1.5	2.6	0.04
Euanthe	1.5	2.6	0.04

Satellite	Mean radius *(km)*	Mean density *(g/cm3)*	Geometric Albedo
Euporie	1.0	2.6	0.04
Orthosie	1.0	2.6	0.04
Sponde	1.0	2.6	0.04
Kale	1.0	2.6	0.04
Pasithee	1.0	2.6	0.04
S/2002 J 1	1.5	2.6	0.04
S/2002 J 1	2.0	2.6	0.04
S/2003 J 2	1.0	2.6	0.04
S/2003 J 3	1.0	2.6	0.04
S/2003 J 4	1.0	2.6	0.04
S/2003 J 5	2.0	2.6	0.04
S/2003 J 6	2.0	2.6	0.04
S/2003 J 7	2.0	2.6	0.04
S/2003 J 8	1.5	2.6	0.04
S/2003 J 9	.05	2.6	0.04
S/2003 J 10	1.0	2.6	0.04
S/2003 J 11	1.0	2.6	0.04
S/2003 J 12	.05	2.6	0.04
S/2003 J 13	1.0	2.6	0.04
S/2003 J 14	1.0	2.6	0.04
S/2003 J 15	1.0	2.6	0.04
S/2003 J 16	1.0	2.6	0.04
S/2003 J 17	1.0	2.6	0.04
S/2003 J 18	1.0	2.6	0.04
S/2003 J 19	1.0	2.6	0.04
S/2003 J 20	1.5	2.6	0.04
S/2003 J 21	1.0	2.6	0.04
S/2003 J 22	1.0	2.6	0.04
S/2003 J 23	1.0	2.6	0.04

(Satellite parameters courtesy of Robert Jacobson, NASA/JPL-Caltech)

CHAPTER 12

 # Saturn: Spectacular Lord of the Rings

In This Chapter

➤ Mystery of the rings revealed

➤ Awesome from any angle

➤ Saturn in the infrared

➤ Winds, storms and a mysterious vortex

➤ Sending spacecraft to Saturn

Mention the planet Saturn and the first word that comes to mind is probably "rings." "Beautiful," "spectacular," and "magnificent" are other words Saturn routinely evokes, even if seen through a very small telescope. If Jupiter is the king of our solar system, Saturn is the "star" (metaphorically speaking) and every star has that something that makes it stand out. Those unmistakable rings make Saturn easy to identify and indisputably striking to look at. Even Venus—dazzlingly bright and named for the goddess of beauty—can't compete with the rings.

Aside from their aesthetic impact, Saturn's rings alter its shape so it does not look round like the other planets. The idea that one planet should look so remarkably different was a puzzle. By improving his telescope lenses, Galileo was able to bring celestial objects into sharper view, but he still had a crude viewing instrument. All he could tell was that Saturn had an odd shape. He had no idea what caused that anomaly. To Galileo, the unusual planet simply looked like it had "ears."

It was not until 1655 that Christiaan Huygens, using superior telescopes, clearly observed a ring around Saturn. He was the first one to realize that the ring was what gave the planet its definitive, not quite round shape. A year before that, Huygens also discovered Titan, Saturn's largest moon, which we'll discuss in the next chapter.

Huygens described Saturn as "surrounded by a thin, flat ring." But what was it made of? Was it solid as it appeared through a small telescope lens? Or was that image deceiving?

Giovanni Domenico Cassini is credited with recognizing that Saturn's "ring" was actually "rings" in the plural. In 1675, he showed that Saturn's ring was composed of numerous smaller rings with gaps in between them. The most notable gap bears his name: the Cassini Division.

Solar System Scoop

Cassini and Huygens share the title of the NASA/ESA spacecraft sent to explore the Saturn system. Cassini-Huygens was launched on October 15, 1997 and began orbiting Saturn on July 1, 2004. The spacecraft's two key components are the Cassini orbiter and the Huygens probe.

The mystery of what Saturn's rings were really composed of would have to wait until the nineteenth century to be solved. James Clerk Maxwell, the great physicist who unified electricity and magnetism, had a lesser known distinction involving Saturn's rings. In the mid-nineteenth century, Maxwell won an undergraduate prize at Cambridge University for mathematically demonstrating that Saturn's rings could not be solid or they would destabilize and break apart. He theorized that Saturn's rings are made up of innumerable objects individually orbiting Saturn. In 1895, spectrographic studies of the rings conducted by James Keeler of the Allegheny Observatory confirmed that Maxwell was right.

The biggest gaps in the rings like the Cassini Division and the Encke Gap are easily seen by telescopes based on Earth. Voyagers 1 and 2 both revealed the intricacies of the rings, with thousands of small gaps and ringlets. Hubble Telescope images bring the Cassini Division into high-definition close-up (Figure 1), and highlight the myriad gaps in the rings.

Figure 1 shows Saturn at its equinox, with sunlight directly hitting its rings. The bands and other features of Saturn's clouds change appearance as the seasons change over the course of a very long Saturn year.

Saturn, with the Cassini division showing clearly in its rings.

The Famous Rings Revealed

De-romanticized, Saturn's spectacular ring is a collection of icy rocks of various sizes and shapes. The biggest ones are about the size of a house and the smallest are no more than particles of dust with individual orbits. Even the most insignificant particles are still bound by Kepler's laws of planetary motion. Therefore, the outer objects (or particles) take a longer time to orbit than the inner objects, in accordance with Kepler's third law.

Saturn orbits the Sun every 29.5 years. Because its axis of rotation always points in the same direction in space, we get to see the planet—and its rings, which circle its equator—from different angles. These variations make viewing Saturn even more impressive. Sometimes the rings appear upraised toward us, sometimes they are tipped downward. And for the most unusual view, we get to see the rings edge-on. Edge-on, the rings are so thin that they virtually disappear (Figure 2).

Figure 2 shows six years of Saturn as seen from a terrestrial telescope. First (lower right), there was Saturn's south polar region. In 2009, our planet passed through the plane of Saturn's orbit. With almost 30 years to orbit the Sun, we won't be going through the plane of its rings again until 2025 and 2039.

Saturn's rings are extremely thin, especially compared to their very wide expanse. The rings are no more than a few tens of meters thick, but they span out for hundreds of thousands of kilometers.

Smart Facts

Saturn Facts

Saturn's orbit

Average distance from Sun	9.5 AU	1,429 million kilometers
Period by the stars	29.4 years	
Period from the Earth	378 days	
Orbit's eccentricity	0.05	
Orbit's tilt	2.5°	

Saturn the planet

Saturn's diameter	120,536 km	9.4 times Earth's
Saturn's mass	95 times Earth's	
Saturn's density	0.7 times water	
Saturn's surface gravity	1.07 Earth's	
Sidereal rotation period	10 hours 40 minutes	

Saturn, like Jupiter, is a giant planet. Its diameter is almost 10 times the diameter of our own planet and it has 830 times Earth's volume. However, unlike our rocky home, its density as a gas giant is very low. Saturn is only 70% as dense as water, while Earth is 5.5 times as dense.

Saturn spins in 10 hours and 40 minutes, less than 10% slower than Jupiter. But Saturn is roughly 15% smaller than giant Jupiter, so its slow rotation makes it even more oblate. Saturn's equatorial diameter exceeds its polar diameter by about 10%.

From Earth. we see Saturn's rings reflected by sunlight. Flying by Saturn, the Pioneer and Voyager spacecraft could glance back and see sunlight scattered forward. Only dust-sized particles scatter light forward like that. Anything bigger, even a pebble-sized particle is opaque. Fortunately, Saturn's rings have a plethora of dust-sized, light-scattering particles for enhanced imaging.

The Cassini spacecraft has been orbiting the Saturn system since 2004. Like the Pioneer and Voyagers, it captures views of Saturn in forward-scattered light, but with far superior resolution. Its images of Saturn show amazing detail (Figure 3). Taken from a high angle above the rings, Cassini captured Saturn's shadow, cast by the Sun. The shadow of the rings is also visible on Saturn's disk. The rings are shown in forward-scattered light.

Solar System Scoop

Saturn is beautiful viewed from even the smallest telescope. A mid-sized telescope reveals the Cassini Division, dividing the rings. And Titan is telescope-friendly. You can even see Saturn's rings with a good (big) pair of binoculars, providing you hold them steady. In brightness, Saturn equals most of the brightest stars, though it is less than half as bright as Sirius.

Saturn is in opposition—opposite to the Sun in the sky and so at its highest point at midnight—in May 2014 and 2015 and June 2016 and 2017. During that time it travels from Virgo to Libra.

Figure 3

Energy in the Infrared

As an object of fascination for hundreds of years, Saturn has gone from the not-quite-round planet with "ears" to a system composed of myriad orbiting objects forming a magnificent ring around a gas giant (not to mention the moons). New technologies are continually expanding the scope of astronomical exploration. The ability to see into the longer wavelengths of light in the infrared gives astronomers a tremendous boost. For planetary objects, the strength of radiation emitted by particles peaks in the infrared. That means the infrared can show energy directly emitted by rocks and ice—a great advantage over relying only on reflected sunlight.

Figure 4

In Figure 4, we see an artist's illustration of an infrared view of a giant super-ring around Saturn. Discovered in 2009, the particles in the ring are so sparse that the ring is not even visible in reflected sunlight, unlike the other rings. Here's where the advantage of infrared comes into play. Given the temperature of the particles, they emit enough radiation at an infrared wavelength roughly seven times longer than red light (5 micrometers), which allowed the Spitzer Space Telescope to detect the elusive ring.

The ring material extends from about 3.7 million miles (6 million kilometers) from Saturn out to 7.4 million miles (12 million kilometers). That's equivalent to 300 Saturn diameters! The inset shows an enlargement compared with the tiny circle showing Saturn's actual size. The image in it is an infrared view of Saturn captured by one of the 33-foot (10-meter) Keck telescopes atop Mauna Kea.

Saturn's bland atmosphere can be seen in Figure 5, which shows the rings almost edge-on. Two of the moons (which we will discuss in a later chapter) are visible in the image. Rhea appears at the right edge of the photograph. As revealed by the phase of Rhea and Saturn, the Sun is off to the left. A shadow of Tethys is (barely) visible near Saturn's left limb, just under the rings.

Saturn gives off roughly twice the energy it receives from the Sun. That implies that it must have some internal energy source. And the proportion of internal energy exceeds giant Jupiter's. Some of it could be residual energy from Saturn's formation, and some degree could come from continued contraction due to a powerful gravitational pull. Helium sinking through the liquid hydrogen in Saturn's interior could be another energy source. Farther away from the Sun, Saturn is colder than Jupiter, which allows helium to condense. Jupiter is the only gas giant on which that does not happen. Saturn, Uranus, and Neptune all share this property.

Figure 5

Saturn's Windy Atmosphere

Saturn's fast rotation means that the clouds in its atmosphere are drawn out into horizontal bands like Jupiter's. But unlike Jupiter, Saturn's atmosphere is distinguished by very high winds. Saturn's surface winds blow over 1,000 miles per hour (1,800 kilometers/hour)—four times faster than Jupiter's winds. Orbiting Saturn, Cassini tracks its winds over time and monitors variations as sunlight hits Saturn at different angles over the course of its changing seasons. The Hubble Telescope also monitors Saturn's winds.

On Jupiter, the various wind zones are fairly well-aligned with the different light and dark bands. This is not quite the case on Saturn, where there is less correlation between the winds and the bands as we see them. Saturn's wind speed peaks at its equator and tapers off at the higher and lower latitudes. There seems to be a lesser peak near the latitudes 50 degrees north and south of the planet's equator.

Nearly twice as far from the Sun as Jupiter, each point on Saturn's surface receives two squared, or four times less energy. That leaves a lot less energy to produce chemical reactions in Saturn's surface, making Saturn's bands a lot less colorful than Jupiter's.

Saturn's Magnetic Field

Even before spacecraft flew by the Saturn system, astronomers were aware that Saturn had a magnetic field. Relatively steady radio signals from the planet alerted them to its presence. A magnetic field is key to the mechanism that gives off radio waves.

Saturn's magnetic field was first measured by the Voyager spacecraft. From near the equator, it turned out to be only two-thirds as powerful as Earth's magnetic field. However, that's still fairly strong when compared to Jupiter's magnetic field, which is stronger than Saturn's by a factor of roughly 20.

Weak or not, Saturn's magnetic field enables it to trap particles from the Sun. These trapped particles generate Saturn's auroras, which can often be seen near the poles (Figure 6). But unlike Jupiter's auroras, which are also observable in visible light, Saturn's auroras are only detectable in the ultraviolet.

Figure 6. Observations by the Hubble Telescope reveal minute-to-minute changes in the auroras. These quick fluctuations are caused by interactions between Saturn's magnetic field and solar wind particles from the Sun.

Stormy Saturn

Cassini has also detected rapidly changing bursts of radio waves coming from a tremendous thunderstorm at Saturn's middle latitudes. The quick radio bursts emanate from lightning caused by static electricity generated by the powerful storm.

In addition, Cassini captured another huge storm towering over the south polar region. The storm spreads 5,000 miles (8,000 kilometers) across, with winds of roughly 300 miles per hour (550 kilometers/hour). At the edge of the storm, there are clouds soaring as high as 45 miles (75 kilometers) above the eye of the storm clouds. Interestingly, this was the first storm with an eye seen on a planet other than Earth. Unlike our terrestrial hurricanes, however, the Saturn storm does not drift.

Even stranger still, there is a vortex around the north pole that appears as a hexagonal shape with fairly sharp corners. And like many things about Saturn it is huge—15,000 miles (25,000 kilometers) across. Solving the mystery of the giant vortex (or what has been solved) was like putting together a puzzle. Since Cassini never flies over Saturn's poles, its shape was configured from piecing together numerous images drawn from Cassini's observations of the lower latitudes.

With relatively high temperature readings, the hexagon shows a clear area that allows the spacecraft's instruments to examine lower altitudes. The scientists were able to discern the shape of the giant vortex but not *why* it should be that shape, or even *how* it can exist. The idea that such a straight-sided figure can even exist is a mystery. On Earth, the closest parallel is a polar vortex that is fairly circular. It is possible that the hexagon is some type of standing wave (a variation that is fixed rather than moving). Or it might even be a novel type of aurora, never before discovered.

Whatever it is, it remains a mystery, and one that intrigues today's scientists the same way their historical counterparts were determined to discover the secret of Saturn's rings. In late April 2013, the mysterious vortex turned up another intriguing phenomenon. Cassini presented scientists with the first close-up visible-light images of a massive hurricane swirling within the hexagonal vortex.

The polar storm surprised scientists by its close resemblance to hurricanes on Earth. But with one major distinction: the monster storm was calculated at *50 times* the size of the average hurricane on our own planet. The storm has probably been swirling for years, but has just been made visible by Cassini's current vantage point, which has a good view of the polar regions. No doubt we'll be seeing more images of Saturn's super storms.

Flying By, Orbiting Saturn

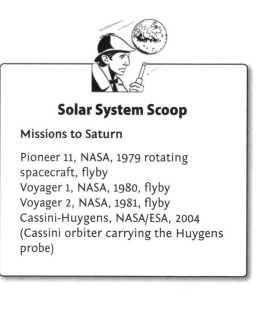

When Pioneer 10 flew by Jupiter in 1979, it was deliberately deflected by Jupiter's gravity to soar out of our solar system without passing any other intrasolar object. Pioneer 11 was different. That mission harnessed Jupiter's gravity to direct the spacecraft toward Saturn. Even if images made by the spinning spacecraft were less than perfect, they provided us with the first close-ups of beautiful Saturn and its amazing rings. Pioneer 11 reached Saturn in 1979. In its long venture, it traveled nearly five years across the solar system from Jupiter.

Solar System Scoop

Missions to Saturn

Pioneer 11, NASA, 1979 rotating spacecraft, flyby
Voyager 1, NASA, 1980, flyby
Voyager 2, NASA, 1981, flyby
Cassini-Huygens, NASA/ESA, 2004 (Cassini orbiter carrying the Huygens probe)

Also with gravitational boosts from Jupiter, the Voyagers aimed for Saturn. Voyager 1 flew by Saturn in 1980, and Voyager 2 followed the next year. Chronologically, the Voyagers were not much later than Pioneer 11, but they far outclassed it in imaging technology.

After arriving in 2004, Cassini completed its primary mission in 2008, and then had its mission extended to 2010 and since then to 2017. Its orbital inclination changes every few years. In 2017, its orbit will be altered by an encounter with Titan, positioning the spacecraft only 1,520 miles (3,000 miles) above the tops of Saturn's clouds. After that, a future encounter with Titan will send the Huygens probe plunging into Saturn's atmosphere.

 # Saturn's Majestic, Magnificent Rings

In This Chapter

➤ Seventeenth century telescopes to twenty-first century spacecraft

➤ Denser main rings, dusty tenuous rings

➤ Tidal forces act on the rings

➤ Close-up views and the sophisticated ring structure

With due respect to our Sun, Saturn's rings are our solar system's most magnificent and impressive feature. No other planet has anything close to Saturn's ring system, spanning out over hundreds of thousands of miles and appearing to us in varying hues.

As we discussed in the preceding chapter, Galileo was the first person to see Saturn's signature feature, but had no idea what it was. Bringing the image into sharper focus, Huygens realized that there was a ring surrounding Saturn, but what he thought he saw was a singular flat disk. Thanks to Giovanni Domenico Cassini, by the end of the seventeenth century, it was known that Saturn's ring was actually made up of multiple smaller rings.

And in the twenty-first century, we have Cassini's namesake spacecraft in Saturn's orbit (along with the Hubble Space Telescope in low Earth orbit) to provide us with unprecedented understanding—and exceptional images—of Saturn's glorious rings.

The rings are lettered alphabetically in the order of their discovery. The main rings are the A, B, and C rings, with rings A and B separated by the Cassini Division. Another famous gap in the rings, the Encke Gap, resides within the A ring. The main rings are denser and composed of larger particles than the faint and tenuous rings made up of dust particles. The B ring is the thickest and brightest of the group. The "dusty rings" include the D ring, extremely faint and in closest proximity to Saturn; the skinny F ring just outside the A ring; the E ring, a faint, attenuated exterior ring; and the G ring, outside the main rings.

An eclipse of the Sun by Saturn, with Saturn with its rings backlighted. The image is a mosaic of 165 different frames taken with the Cassini spacecraft. Unlike an eclipse of the Sun on Earth by the Moon, the Sun is much smaller than Saturn's disk as seen from the Cassini spacecraft with the Sun partway into the disk, and we see a bit of its brightest light refracted (bent) by one location near Saturn's southern pole. The dark side of Saturn, which faces up and which is seen from an angle of 15°, is faintly lit by reflections by Saturn's own ring.

The rings are very bright and (by astronomical standards) very young. Their remarkable brightness led scientists to believe in their youth. Had the rings been around for the billions of years since the solar system was born, they would have been covered in dust and much darker. Additionally, calculations of how the gravity of the moons interacted with the rings point to the idea that the rings are no more than a few hundred million years old.

The rings are composed primarily of water ice with traces of rocky material. The material is replenished over time, probably from a moon breaking up under the onslaught of relentless impacts and collisions. The main rings all differ slightly in color, which may mean that each of the rings came from bits of a different shattered moon.

Tidal Forces at Work

We've already gone over the tidal forces on Earth, how the gravity from the Moon (and to a lesser degree, the Sun) cause the tides on our earthly oceans. As a differential force, the tides are produced by the difference between the force on one side of the planet and the force on the other side. The more powerful gravitational pull on the side nearest the Moon causes the difference in the two forces.

When an object is too close to its parent planet, the differential force is so powerful that it makes it impossible for small bodies to form a moon. Larger moons formed from small bodies eons ago are easily ripped apart by the ultra-powerful gravity, as are any liquid or molten bodies. As a result of this phenomenon, there is a zone known as the "Roche lobe" (for the nineteenth-century French astronomer Édouard Roche) inhabited by rings rather than moons. In the Saturn system, for example, the moons exist almost entirely outside the rings.

Only artificial satellites, composed of metal and other sturdy materials, can withstand the force of gravity inside the Roche lobe. Natural agglomerations of materials are no match for the powerful gravitational pull. The rings around Uranus and Neptune, which we'll learn about later, like the rings of Jupiter and Saturn, are the product of tidal forces from their parent planets.

Saturn's Rings at Close Range

Flying by Saturn, Voyagers 1 and 2 showed the major rings breaking up into hundreds of ringlets. First the Voyagers, and then Cassini, captured the shadow of the rings on Saturn's surface. Thanks to photographs taken in forward-scattered light, we now know that the Cassini Division and Encke Gap are both filled with smaller particles than the icy rocks or boulders easily visible in reflected sunlight. The thickest parts of the rings are opaque (Figure 2).

We look through the rings to see sunlight scattered toward us. In this view from the Cassini mission, the dark parts are the thickest parts of the ring. Some small storms are visible at higher latitudes on Saturn's surface.

Intricacies of the Ring Structure

Apart from the powerful tidal force that created them, the primary force acting on Saturn's rings is the gravity of the numerous orbiting moons. The legions of ringlets are held in place by moonlets, which keep increasing in numbers as new discoveries are made.

Solar System Scoop

These very small moons are sometimes called "ringmoons." Gravitational tugs from the ringmoons keep the ringlets from straying. Gravity from a ringmoon orbiting slightly outside a ring accelerates the speed of the ring material, which forces it back down and holds it in place. A ringmoon orbiting slightly inside a ring has the opposite effect. Thus, a pair of ringmoons combined can shepherd the ring material, earning them the name "shepherding satellites."

Detailed images of some of the ringmoons have been created from Cassini observations. From these images, scientists have determined their size. They have also determined their masses by deducing them from waves rippling the rings caused by their moons. Together, the sizes and masses give us their density, which turns out to be very low. In fact, the density of the ringmoons is so low—roughly half the density of solid ice—that they are probably nothing more than collections of loose debris as opposed to solid blocks of ice and rock.

Each individual particle in Saturn's rings orbits the planet on an essentially elliptical path. Scientists had expected all of the structure within the ring system to be elliptical or round, like the Cassini Division and Encke Gap. They were amazed when Voyager detected non-circular structure in the form of radial "spokes" (Figure 3).

Figure 3

Cassini's close-up of Saturn's B ring shows, along with the numerous ringlets, spokes of brightness going radially outward. Saturn's gravity alone cannot account for the odd phenomenon. Theoretical explanations centered on static electricity. Specifically, it was proposed that static electricity elevates some of the particles above and below the main ring level. The elevated, electrically charged dust appears bright from forward scattering and dark from back scattering. This explanation has since been validated by Cassini observations of the spokes from different angles.

The spokes are produced by interactions between the ring material and the Sun's light. They form as the ring material emerges from Saturn's shadow and disappear when that part of the ring reverts back into shadow. Cassini has made fascinating movies showing the spokes forming, revolving, and disappearing in remarkable detail.

Some of the rings are relatively isolated from the main rings, which allows their interactions with the ringmoons to stand out. The intricate twisting and braiding of the delicate F ring is especially intriguing (Figure 4). The pattern follows the positions of the orbiting ringmoons, with Prometheus being the major player.

Figure 4

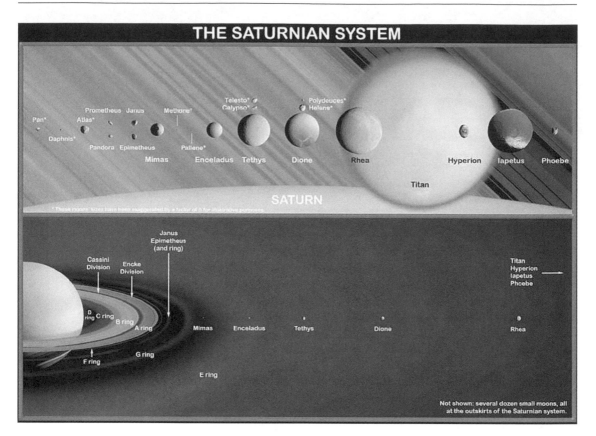

Figure 5

Due to Saturn's Roche lobe, the various rings lie inside the numerous moons (Figure 5). Figure 5 provides an excellent picture of the structure of Saturn's rings and the relative positions and sizes (to scale) of the biggest moons of the Saturn system. Note the huge scale of the gas giant itself.

In the next three chapters, we will be covering the most interesting of Saturn's moons. No Galilean quartet, but at 63 moons, Saturn has almost as many as Jupiter.

CHAPTER 14

Saturn's Titan: The Moon with Smog and More Smog

In This Chapter

➤ Titan, Saturn's largest moon

➤ Titan's thick, smog-filled atmosphere

➤ Clouds, rain, and an obscure surface

➤ The Huygens probe lands on Titan

➤ Cassini picks up the search for lakes

The Dutch astronomer and physicist Christiaan Huygens was an unofficial protégé of Galileo. Inspired by Galileo's discovery of the four large Jovian moons, and his work in improving telescope capabilities, Huygens enlisted his brother Constantijhn in building their own telescopes. Titan came into view in 1654, a year before Christiaan's observations of Saturn's ring. In 1655, Christiaan Huygens published *De Saturni Luna Observatio Nova*, chronicling his discovery of *Saturni Luna* (or *Luna Saturni*, "Saturn's moon"), which we now know as Titan.

Titan, Saturn's largest moon, has a thick, smog-filled atmosphere composed mainly of nitrogen. It shares that distinction with one other object in our solar system: our home planet Earth. This resemblance to Earth is a powerful driver of missions to find out as much about Titan as we possibly can. In some respects, giant Titan, orbiting the ringed gas giant Saturn, is actually more like Earth than other place in our realm of space. Titan has wind and rain and seasonal weather patterns like Earth, and as a result, it has familiar surface features such as dunes, deltas, seas, rivers, and lakes. Though at an average surface temperature of about -290.56 degrees F (-179.2 degrees C), it is less than hospitable to human life.

Titan can be seen through a small telescope or even strong binoculars. However, it can be tricky to find because it is so close to dazzling Saturn and the ring system. Seeing Titan in *detail* is especially challenging due to its dense, almost impenetrable atmosphere.

Solar System Scoop

The names we use for the Saturnian moons were the idea of John Herschel. In 1847, Herschel proposed naming the moons after mythological figures associated with Saturn, the Roman god of agriculture and harvest (in Greek mythology, Cronus). The seven moons known in Herschel's day are all named for brothers and sisters of Saturn, all members of the powerful race of Titans, descendents of Gaia (Mother Earth) and Uranus (Father Sky).

Titan himself was Saturn's older brother, but was prevented by Saturn from assuming the throne. In fact, Saturn jealously guarded his status, to the point that he ate his children to prevent them from usurping his throne. Galileo evoked this myth when in 1612 the plane of the rings faced directly toward Earth, making the rings invisible to his view. "Has Saturn swallowed his children?" exclaimed the puzzled Galileo. He was even more mystified when the rings reappeared in 1613.

NASA and ESA have both invested heavily in missions designed to expand our knowledge of Titan. Unlike Galileo's Jupiter probe, which went into the planet itself, the Huygens probe went through Titan's thick atmosphere to land on the intriguing moon. The probe landed six and one half months after the Cassini-Huygens spacecraft began its orbit around Saturn.

In Cassini's visible light image of Saturn (Figure 1), we can see Titan's reddish, smoggy atmosphere—but not its surface, which is obscured by the smog. The image provides a good perspective of Titan's size by juxtaposing Titan (behind the rings) with the much smaller Epimetheus (above the rings). Titan spans 3,200 miles (5,150 kilometers) across, compared to Epimetheus's 72 miles (116 kilometers). Titan is more than 40 times the size of Epimetheus.

The dark area in the main ring is the Encke Gap. The tiny moon Pan, only 16 miles (26 kilometers) across, is visible in the Encke Gap.

Smart Facts

Titan in Perspective

	Diameter	
Moon (Earth's)	3,476 km	27% Earth's
Mercury	4,880 km	38% Earth's
Titan (without its atmosphere)	**5,150 km**	**40% Earth's**
Ganymede	5,268 km	41% Earth's
Mars	6,794 km	53% Earth's
Earth	12,756 km	

The True Titan

Even before spacecraft ventured toward the Saturn system, astronomers were aware of Titan's thick, opaque atmosphere from terrestrial images, polarization measurements, and spectra. The Voyager images were somewhat disappointing. They showed no discernible features. Titan was still a mysterious object, but one definitely worthy of future exploration.

A weakening of the radio signal from Voyager 1 provided a clue to Titan's atmospheric pressure. When the spacecraft flew behind Titan, its signal faded in a particular way. By analyzing that fade, scientists were able to calculate that the atmospheric surface pressure right above Titan's surface is 50% higher than the atmospheric surface pressure on Earth.

Titan's signature reddish tinge is caused by the smog—a familiar occurrence to residents of many terrestrial cities. (Los Angeles tops the chart as the smoggiest U.S. city, but pales in comparison to other cities such as Beijing.) Sunlight acting on the chemicals in Titan's atmosphere produces the color and the opacity.

The Voyagers captured several layers of haze at Titan's limb. They also discovered the nitrogen and revealed that nitrogen (in the form of N2, the nitrogen molecule) accounted for most of Titan's atmosphere, making it similar to our own familiar atmosphere. Methane accounts for less than 1% of Titan's atmospheric gas.

Solar System Scoop

The discovery of Titan's soupy atmosphere actually *demoted* Titan, which held the status of the largest natural satellite in the solar system. Without its atmosphere, Titan turns out to be slightly smaller than Ganymede, which has virtually no atmosphere.

As frigid as Titan is, it is actually warmer than we would expect for an object so far from the Sun. Titan receives a scant 1% of the sunlight we enjoy on Earth. But in one of its many resemblances to Earth, Titan is also home to a greenhouse effect.

There are actually two forces affecting Titan's equilibrium temperature. One is the greenhouse effect, caused by the methane gas. It is even possible that the smoggy, icy moon was once warm enough to support life. At the same time, the smog blocks out some of the light from the Sun, for a cooling effect.

Based on Voyager measurements, the two effects created about 18 degrees F (12 degrees C) of warming beyond the expected equilibrium temperature. The Voyagers' infrared radiometers recorded the temperature of-290.56 degrees F (-179.2 degrees C), or 93 degrees C above absolute zero.

On Earth, we have abundant water in all three of its phases: solid (ice), liquid (water), and gas (water vapor). At normal surface pressure, the change in form occurs around 32 degrees F (0 degrees C), a temperature known as the "triple point." Titan's measured temperature is near the triple point for methane. This suggests that methane plays a similar role on Titan to the role of water on Earth.

The Voyager readings prompted the idea that Titan was home to lakes of methane or ethane. Interesting as it was, proving it was a daunting challenge, in view of Titan's impenetrable atmosphere. Though as with many theories, proof would eventually come after a wait, in this case three decades (more of that in this chapter).

Infrared light was able to penetrate the dense clouds to some degree. The Hubble Space Telescope's infrared detector discerned some surface structure on Titan, but nothing that could be defined as a lake.

Cutting-edge adaptive optics for ground-based telescopes proved a great boon to scientists striving to see features on Titan's surface. Peering through the infrared, scientists using the Keck II telescope at the Mauna Kea Observatory were able to make out more detail on Titan. By penetrating the clouds, the infrared waves revealed a bright continent, along with a few methane clouds near the satellite's south pole.

Due to Titan's distance from Earth, radio waves are less successful, even though they are capable of penetrating the thick clouds. Titan is almost 10 AU from the Sun, and consequently never closer than about 8.5 AU from Earth (compare that to the 0.3 AU that sometimes separates Venus and Earth). At that great distance, even the radar from the ultra-powerful Arecibo radio telescope was unable to provide details of Titan.

There is no question that Titan presented a challenge, even with the most advanced technologies. Scientists have long been determined to discover the secrets of Titan's surface because conditions like smog and lightning signify that organic (carbon-containing) materials probably rained down on the Saturnian moon. Though Titan is obviously much colder than Earth, the overall picture suggests that conditions on Titan may be similar to Earth as it was eons ago when life began. Detailed study of what exists under Titan's thick clouds holds tremendous fascination for anyone interested in the origins of life.

Solar System Scoop

The rain from Titan's clouds was confirmed by observations from a number of terrestrial telescopes on Mauna Kea. The list includes the Keck Telescopes of the University of California and the California Institute of Technology, the Canada-France-Hawaii Telescope, and the United Kingdom Infrared Telescope.

Titan is the only natural satellite with a thick atmosphere. Its distance away from the Sun may be the reason. Titan is almost twice as far from the Sun as Ganymede and Callisto, which have similar surface gravities. Solar heating, combined with additional warming from Jupiter, would have caused any early atmosphere in its moons to escape. Far out in the solar system, Titan would not have been subject to the same effect as the Jovian moons.

Landing on Titan

In view of scientists' intense fascination with Titan, it is not surprising that the Cassini-Huygens mission keeps being extended. Since 2004, the Cassini orbiter has been sending back spectacular images, aesthetically and scientifically. Titan's value to planetary scientists is evident in the decision to have the probe land on the moon instead of its parent planet.

Titan shown silhouetted against Saturn's disk. The smog in Titan's atmosphere prevents us from seeing into its surface. Saturn's rings are seen nearly edge-on in this image.

The Huygens probe was launched into Titan's atmosphere on January 14, 2005. The first thing it detected was high winds, gusting up to 250 miles per hour (400 kilometers/hour). Interestingly, the winds were actually measured by terrestrial telescopes, which detected and analyzed the probe's shifting radio frequencies. The equipment aboard the probe was set up to measure velocities by means of the Doppler effect of shifting frequencies.

Huygens was busy measuring the chemical content of Titan's atmosphere throughout its two-and-one-half-hour descent. As it descended farther, the less opaque atmosphere allowed it to capture photographs. The images seem to show a smooth area separated from a rough area by something resembling a shoreline. Channels in the rougher part might have meant that liquid was flowing downward (or did at some time).

Still, there was no more definitive sign of liquid, such as light reflected off a glassy surface that might be a liquid body.

On Earth, scientists, engineers, and an excited general public eagerly waited for radio signals from Huygens. Given the distance of the Saturn system, that meant the wait was at least an hour. Huygens was equipped with a metal rod at its bottom to sense when it reached the surface. The rod broke through Titan's terrain and the space probe settled onto the giant moon to loud, excited cheers. No other spacecraft had ever landed on such a distant object before.

As Huygens landed, a sensor detected a puff of methane. The probe's arrival might have turned methane frost into vapor. Once established on Titan's surface, Huygens engaged in its task of imaging its surroundings (Figure 3). Through the orangey haze, the photographs revealed what appeared to be rocks, probably blocks of ice and hydrocarbons eroded by flowing liquid (methane or ethane). In other words, Huygens found more evidence of a shoreline.

The Huygens lander on Saturn's moon. Titan took this view after its 2½-hour descent in one of the most exciting pictures ever made in the space program. There are rocks made of frozen water and hydrocarbons at the surface temperature of –290°F (–180°C). The batteries lasted over 90 minutes.

Huygens took pictures for more than 90 minutes before it died. It was a triumph that it landed successfully and lasted that long on the surface of the very distant, windswept, smoggy satellite.

The only disappointment is that the probe did not discover liquid on Titan. Despite signs of shoreline and channels, there was no definitive sign of a liquid lake or sea.

Cassini Takes Over

If Huygens didn't find liquid, the Cassini spacecraft was still orbiting Titan, equipped with radar that could penetrate Titan's dense clouds. As the radar mapped large expanses of Titan's surface, it revealed features that looked like lakes surrounded by interesting varied terrain (Figure 4a,b,c). These landscapes were most common at high southern and northern latitudes where liquid would be least likely to evaporate. From the perspective of Titan's size, these lakes are about the equivalent of the Mediterranean Ocean, the Black Sea, or the Great Lakes.

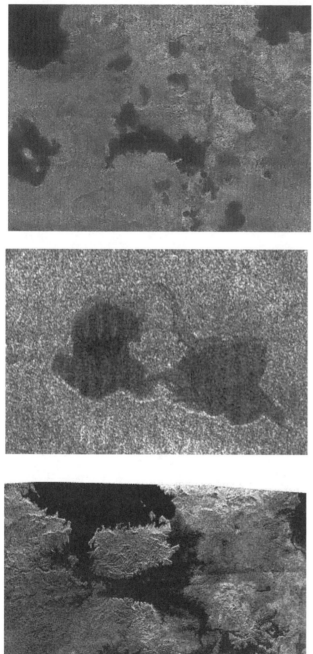

Figure 4a

Figure 4b

Figure 4c

A mysterious dark feature near Titan's south pole was given the name Ontario Lacus. A flyby on July 22, 2006 (during winter at Titan's northern latitudes), captured several smooth, dark spots near the north pole. By January 2007, scientists involved with the mission were convinced that Titan had methane lakes. In June 2008, Cassini's Visual and Infrared Mapping Spectrometer (VIMS) confirmed that Ontario Lacus held liquid methane. By December that same year, a *specular reflection* (a mirror-like reflection of light or another wave from a surface) corroborated the presence of liquid in Ontario Lacus as Cassini flew directly over the lake.

The most definitive evidence appeared on July 8, 2009, when the VIMS caught a specular reflection off a smooth, glassy surface now known as Jingpo Lacus. Located in the north polar region, the lake was just coming out of the darkness of Titan's 15-year winter. As the angle from the orbiter to the lake to the Sun changed, the gleam came into view (Figure 5).

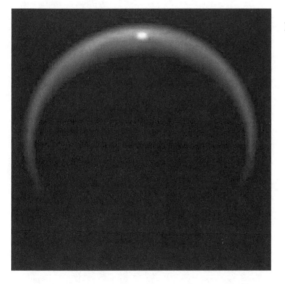

Figure 5

Then, in the course of a flyby on September 26, 2012, Cassini's radar detected what may be a *river* in Titan's north polar region, extending more than 400 kilometers (248.5 miles) and ending in Ligeia Mare. Though it is much smaller in size, the river has been compared to the Nile River on Earth.

It took nearly 30 years from the Voyager indications of methane or ethane lakes on Titan to Cassini's confirmation that the lakes do exist. The Titan Saturn System Mission (TSSM) a joint NASA/ESA venture exploring Saturn, Titan, and Enceladus was originally proposed for launch in 2020. The launch date has since been postponed to some later point in the decade and it will take nine years to reach its destination. Cassini remains in orbit around the Saturn system, sending back images of its inhabitants.

CHAPTER 15

 Saturn's Enceladus: Moon of Many Geysers

In This Chapter

- ➤ An unexpected discovery: huge jets of water ice
- ➤ Imaging and analyzing the jets
- ➤ "Tiger stripes," where the jets originate (and unusually warm)
- ➤ Top candidate in the search for extraterrestrial life

Enceladus, Saturn's sixth largest moon, was discovered by William Herschel on August 28, 1789, with what was then the world's largest telescope. He originally observed it two years earlier, but his smaller telescope was unable to make out the faint celestial object so close to brilliant Saturn. Like the other Saturnian moons, Enceladus is named for one of the Titans (based on the system created by its discoverer's son). The mythological Enceladus was a giant; the moon does not quite measure up. It is only 314 miles (505 kilometers) across, and 10% of Titan's diameter (and thus 0.1% of its volume). But what the mid-sized moon lacks in size, it has features that make it a top priority for astronomical exploration.

Specifically, Enceladus is home to cryovolcanoes spurting out huge jets of water ice. In 2005, a series of close flybys by Cassini produced a truly astonishing discovery: liquid plumes at the south polar region shooting water ice out into space for hundreds of miles. The jets seemed to come from liquid water beneath the moon's icy veneer. This discovery vaulted Enceladus way up on the list in the search for extraterrestrial life.

Since this marvelous find, the path of the Cassini orbiter has been redirected to focus greater attention to Enceladus. And although we will have to wait more than a decade, the Titan Saturn System Mission is targeted to Titan *and* Enceladus.

Exploring Enceladus, Finding the Jets

Before the Voyager missions, knowledge of what Enceladus might look like in detail had scarcely improved since Herschel discovered the small speck near Saturn. Images captured by Voyager 1 in late 1980 were very poor resolution, showing nothing more than a smooth, glossy surface with no impact craters. Closer observation by Voyager 2 in August 1981, at much higher resolution, provided a much better, more detailed picture of the elusive moon. The images reaffirmed that much of the surface is smooth (and therefore, fairly young), but they also picked up cratering of various degrees that Voyager 1 had missed. At mid- to high-northern latitudes, Enceladus is heavily marked by deep impact craters. Lighter cratering dots the area close to the equator. On the southern half of the moon, tectonic activity has erased most of the craters.

Scientists had not expected the degree of geological variation they found on Enceladus. In particular, they were mystified by the youthful terrain—implying geological activity—on such a small, frigid object. With the mystery unresolved, Enceladus became a prime target for exploration by the Cassini-Huygens mission.

Solar System Scoop

In a departure from Greek and Roman mythology, features on Enceladus are given names from *The Arabian Nights*. For example, Julnar, a 20-kilometer-wide crater, is named for a heroine of the *Arabian Nights* tales. The odd Anbar Fossae depressions curving across the moon's surface (Figure 1) bear the same name as Anbar in modern Iraq.

Cassini was programmed for several flybys of Enceladus as well as numerous imaging and data-gathering encounters within 100,000 kilometers (62,500 miles) of the moon. In an unexpected discovery, the spacecraft's Visual and Infrared Mapping Spectrometer detected unusual warmth in a region near the south pole. Analysis of its surface materials revealed an area made up of cracks that seemed to be covered by substances different from anything found anywhere else on Enceladus. The cracks per se seemed improbably warm.

Features on Enceladus are given names from *The Arabian Nights*.

Inspired by the curious temperature anomaly, the Cassini team decided to photograph Enceladus by looking back past the moon toward the general direction of the Sun. It produced a spectacular find: the images revealed about a dozen jets spraying what seemed to be fine, icy particles far out into space (Figure 2). The plumes extended hundreds of miles beyond Enceladus. The discovery was reminiscent of the volcanoes erupting on Jupiter's Io. And Enceladus had seemed like such a nondescript satellite!

Figure 2

Mapping Enceladus allowed the scientists to determine where the jets were coming from (Figure 3). They originate from fractures in the moon's crust known as *sulci,* and colloquially called "tiger stripes." (In Latin, the term *sulci* relates to plowing and means parallel rows or furrows.)

For some time, Enceladus was thought to be the source of Saturn's E ring. The ice particles thrown off by the jets seem to fill a doughnut of material around Saturn, creating the E ring.

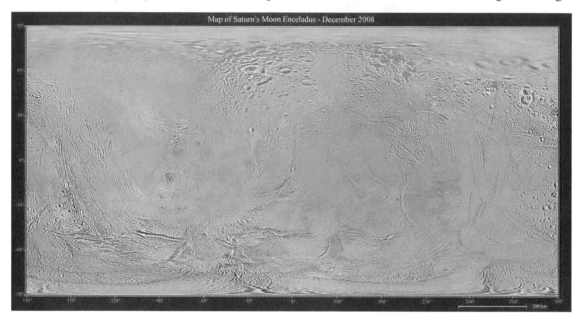

Figure 3

Solar System Scoop

Cassini's spectrometer detected water vapor and an admixture of some organic compounds in the plumes spraying from Enceladus. Some salt has also been found in the plumes. Further analyses of the plumes' constituents have been made by terrestrial telescopes.

The tiger stripes appear to originate from soft ice in the cracks swelling upward. Detailed images reveal areas of Enceladus's crust that have slid along the fractures and have been filled in by new material erupting in the middle. The tiger stripes occupy an area virtually devoid of impact craters.

Modeling the Jets

Several models have been proposed to explain how the jets form (Figure 4). The Cassini Imaging Team Scientists believe the jets are actually geysers erupting from pockets of liquid under the surface of Enceladus. In view of the presence of liquid water, extra warmth, and organic materials, this conception of how the jets form makes Enceladus a prime candidate in the search for extraterrestrial life. Remember, the water believed to be on Europa lies trapped below a very thick layer of surface ice. Liquid water close to the surface exalts Enceladus over Europa in the quest for conditions that might support the origins of life beyond Earth.

A competing model proposes that what is really happening on Enceladus is that warm ice is evaporating and then freezing. The implications of this model are much less exciting. With Enceladus a major target for further exploration, we will eventually find the most accurate explanation.

Plume Vent Models

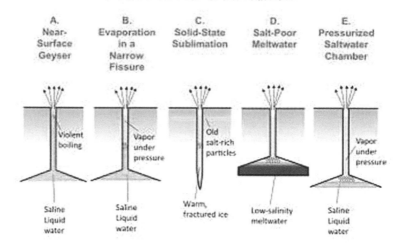

In their suggested caption, the Cassini Imaging Team describes these models for Enceladus's jets as follows:

"These illustrations indicate possible ways in which the water vapor and ice particles in the plume of Enceladus may be formed. The Cassini spacecraft recently found a small fraction of salt-rich ice particles from the plumes, while Earth-based observations indicate gas from the plumes is very poor in sodium. These measurements are helping scientists to evaluate how the plumes form.

"In model A, salty water boils explosively near the surface of Enceladus when it encounters the vacuum of space. This model can be ruled out, because such explosive activity would spread large amounts of sodium into space where it would have been seen by the Earth-based observers. If this model was correct, then nearly all the ice particles observed by Cassini would be salt-rich, instead of just a fraction of them.

"In model B, salty water evaporates more slowly at some depth in a narrow fissure, creating vapor which escapes to the surface to form the plume. This model also seems unlikely because the fissure would rapidly become clogged by salt left behind as water evaporates. The water would also freeze, because not enough heat could reach the water surface up the narrow fissure to replace the heat lost by evaporation.

"In model C, the warm ice evaporates directly into vapor to form the plume, in a process called sublimation. The salty particles found in the plume would have been created by liquid water in an earlier epoch and would have been stored in the near-surface layers of Enceladus until the present. These particles would now be incorporated into the plume by the escaping gases. This model cannot be ruled out, but seems unlikely because it may be difficult to dislodge old ice grains from the walls of the fracture.

"In model D, the liquid water results from melting of near-surface ice rather than coming from an underlying salty ocean. The water is initially only slightly salty, but its salinity increases as evaporation removes some of the water and leaves the salt behind. Thus, in this model, the salt-rich ice particles seen by Cassini would be derived from initially salt-poor water. This model may be plausible and has not yet been evaluated in detail.

"In model E, the water is originally salty, and perhaps comes from a subsurface ocean in contact with an underlying rocky core. The water evaporates slowly into a pressurized chamber, from which water vapor and ice particles, including salty particles from the salt water, escape to the surface along narrow fissures. The large area of the evaporating water surface prevents accumulated salt from clogging the vent and allows enough heat to reach the water surface from below to prevent the water from freezing. This model seems the simplest, and perhaps most likely of the models shown here, but is not the only possibility. Enceladus' plumes may involve a combination of several of these idealized models."

Tiger Stripes

Cassini has recorded temperatures of -154 degrees F (-103 degrees C) near the tiger stripe known as Damascus Sulcus: more than 180 degrees F (100 degrees C) warmer than the surrounding terrain (Figure 5). Temperatures closer to the fractures may be even higher, and it may be warmer still under the surface. The exciting implication of this phenomenon is that the temperature underground gets high enough to melt ice and create liquid water. That model goes well with Cassini's discovery of salt in the icy plumes.

Figure 5

In Figure 5, we have a false perspective view of Damascus Sulcus, calculated from Cassini observations made at a resolution of 40 to 100 feet (12 to 30 meters) and an overall map of Enceladus. To bring out the structures, the altitudes were exaggerated by about 10 times.

Damascus Sulcus is composed of two large parallel ridges with a ditch 650 to 820 feet (200 to 250 meters) between them, for a total width of three miles (five kilometers). Each ridge is 325 to 500 feet (100 to 150 meters) high. The ditch was probably formed by tidal forces shearing and sliding the crust.

The ridges surrounding the tiger stripes are composed of much coarser ice than the fine-grained ice coating most of Enceladus's terrain. The tiger stripes vary in geological activity. Damascus and Baghdad sulci are the most active and Alexandria Sulcus the least active.

Enceladus Close-Up

From the planning stages of the Cassini-Huygens mission, Enceladus was targeted for close flybys. The exciting implications of liquid water on its surface intensified the drive to know more about Enceladus. Tremendous efforts were made to see that the spacecraft took numerous close-up shots of the Saturnian moon. Close-up views of the jets can be astonishing, showing images such as the jets erupting tens of meters, feeding the bigger plumes (Figure 6).

Cassini has even flown right through some of the jets, directly sampling the particles.

In May 2011, NASA scientists attending an Enceladus Focus Group Conference declared that Enceladus "is emerging as the most habitable spot beyond Earth in the Solar System for life as we know it." Enceladus has dethroned Europa as the top contender in the search for extraterrestrial life.

CHAPTER 16

Saturn's Moons: Rivaling Jupiter

In This Chapter

> ➤ Many interesting moons in the Saturn system
> ➤ Mimas, the moon with a death star crater
> ➤ Tethys, home to a giant crater and giant canyon
> ➤ Dione, Rhea, Hyperion, Iapetus
> ➤ Discovering more moons

From the time Galileo first noticed the "ears" on Saturn—and was astonished to find that they disappeared and then reappeared again from his view—Saturn has been a constant source of fascinating discoveries. The expansive signature rings surrounding the planet, the giant moon Titan, and the geysers of water ice spraying out hundreds of miles into space from modest-sized Enceladus are only some of the numerous objects and attributes that make the Saturn system one of the most interesting spots in our solar system.

Saturn has 62 moons, only a few satellites short of Jupiter's 67. Eight of those moons were discovered by direct observations from telescopes based on Earth. We've already discussed two of them, Titan, the biggest, and Enceladus, the most exciting for learning the origins of extraterrestrial life. Mimas, Tethys, Dione, Rhea, Hyperion, and Iapetus are the other six major Saturnian moons. We will cover them in this chapter.

Fifty-three of the moons have names. The remaining moons have only been seen once and will have to be seen again before they earn names. Since the rings are inside Saturn's Roche lobes, all of these moons orbit outside the rings. The tiny ringmoons, as we learned, control the ring structure and shepherd the narrow ringlets.

The very small size of most Saturnian moons is highlighted by the fact that Titan accounts for more than 96% of the mass orbiting Saturn. Six of the seven additional major moons (with the exception of small, irregularly shaped Hyperion) make up roughly 4% of the mass, and the remaining moons and the rings account for a scant 0.04%.

All the moons we'll discuss in this chapter are low in density—only one to one-and-one-half times the density of water and one-fifth the density of the rocky terrestrial planets. Consequently, they must be composed mainly of water ice with bits of rocky material. In an unusual twist, there is no relationship between their densities and their proximity to Saturn. All these small moons, remote from the Sun, share hard, cold, icy surfaces that retain impact craters from all those early collisions.

Smart Facts

The Major Saturnian Moons

	Diameter
Mimas	400 kilometers
Enceladus	500 kilometers
Tethys	1,054 kilometers
Dione	1,120 kilometers
Rhea	1,528 kilometers
Titan	5,150 kilometers
Hyperion	360x280x226 kilometers
Iapetus	1,436 kilometers

Like just about everything around Saturn, these moons have unique and unusual features. The largest of the moons orbit with one side gravitationally locked toward the parent planet. Hyperion and Iapetus are two of the strangest satellites in our solar system. Janus and Epimetheus share the same orbit and sometimes change places. There are a lot of marvelous goings on in the Saturn system.

Mimas

Medium-sized for a small moon, Mimas has a radius of roughly 200 kilometers. It ranks as the smallest of the inner round moons, though measurements from space reveal it to be not quite so round.

Mimas is distinguished by a gigantic impact crater that makes the moon resemble the death star from *Star Wars*. Spanning 130 kilometers across (for perspective, Mimas's diameter is only 400 kilometers). The crater is named Herschel for William Herschel, who discovered Mimas and Enceladus (and Uranus, as we will get to in the next chapter). The raised rim and central rebound of the huge crater are similar to some of the craters on our own Moon (Figure 1a).

Figure 1a

On the other side of the satellite, halfway around from Herschel, is a great canyon (Figure 1b). It probably formed as a result of the massive impact that formed Herschel and shattered part of the small moon. Energy from the impact could have been directed by Mimas's surface toward the point where the canyon formed. The overall surface of Mimas is marked by impact craters.

Figure 1b

Tethys

The third largest of Saturn's inner moons, Tethys is dominated by the huge crater Odysseus (protagonist of Homer's *Odyssey*). Odysseus (Figure 2a) measures 450 kilometers (280 miles in diameter)—bigger than Mimas and colossal for an object of Tethys's size.

Figure 2a

Tethys has much more varied terrain than crater-covered Mimas. The variations in Tethysis's surface suggest that it must have been very internally active eons ago before it solidified.

Like Mimas, Tethys has a giant canyon on one side (Figure 2b). Located on the side facing Saturn, Ithaca Chasma covers 620 miles (1,000 kilometers), two-thirds of the way around Tethys. The vast canyon is several miles deep. It might have been formed as Tethys expanded when its interior froze billions of years ago. Another possible explanation is that the canyon is a geological fault that emerged in the frozen ice.

Figure 2b

Ithaca Chasma and Odysseus are concentric and the two features may be related. In *The Odyssey*, Ithaca was Odysseus's home.

Tethys has numerous impact craters, even visible in Ithaca Chasma. Most of its surface is heavily cratered and hilly. A smaller, flatter area on the side opposite Odysseus has fewer craters. There is a clear boundary between the more and less cratered terrains. The presence of craters in the canyon shows that it formed billions of years ago.

Dione

The second-largest inner moon, Dione has interesting variations in its terrain. Though most of its surface is marked by ancient, heavily cratered terrain, there are also bright veins of material that might be signs of internal eruptions (Figure 3). This "wispy terrain" is on the side of Dione that trails as the satellite orbits Saturn and shows material that reached the surface at a relatively recent point in the moon's evolution.

Along with the ubiquitous deep impact craters, Dione's varied surface includes an extensive network of ridges and valleys. It also has some smooth plains with shallow craters that were probably resurfaced by fairly recent tectonic activity. Dione might still be geologically active to some degree, but whether that is the case is a mystery.

Figure 3

Dione has one particularly strange phenomenon for our solar system, even given the odd happenings in the Saturn system. It has a smaller moon sharing its orbit. This "Trojan" moon, Helene, is tiny, only 20 miles (32 kilometers) across. Given names from the Trojan War, Trojan satellites in the Jovian system orbit in stable gravitational positions 60 degrees ahead of and behind Jupiter. Helene assumes a similar Trojan position in relation to Dione.

Rhea

Second in size to Titan among the Saturnian moons, Rhea has craters extending 300 kilometers in diameter. Some of the craters have sharp rims, implying that they were formed more recently than the ancient craters with smooth and eroded rims. Similar to Dione, Rhea shows wisps of light material on its trailing surface, probably the result of internal activity (Figure 4).

Figure 4

In 2005, Cassini's magnetometer detected a blockage in the electron flow around Saturn caused by an area around Rhea. The source was thought to be dust-size particles forming faint rings around Rhea's equator. That would be one more exciting discovery in the Saturn system. Rhea would be the only moon in our solar system to have rings. However, subsequent observations of the hypothesized ring plane taken from various angles by Cassini's narrow-angle camera showed no evidence of the anticipated ring material. What disrupted the electron flow is still a mystery and at this point, Rhea does not appear to have rings.

Hyperion

Hyperion, the smallest of the major moons, is odd in many respects. Discovered in 1848, it was long recognized that Hyperion was much too small to have strong enough gravity to make it round. What no one expected was that the small moon would be shaped like a hockey puck. Images from the Voyager mission provided that unexpected discovery.

Note that Hyperion's diameter is measured by three figures instead of one. It is the biggest moon in the solar system known to have an irregular shape.

Hyperion is close to giant Titan, keeping it at the mercy of Titan's powerful gravity. In fact, Hyperion is affected by an interaction of gravitational forces, with Titan's the dominant force. As a result, it spins chaotically, even abruptly changing direction. No other moon we know of shares its chaotic rotation.

Extremely detailed images from Cassini's cameras revealed even more of Hyperion's strange attributes (Figure 5). For one thing, its icy surface looks like a sponge. Battered by massive impacts over billions of years, Hyperion's surface is marked by countless deep pockets. The moon's low density may make its gravity so weak that its surface is very porous. With very low gravity, material thrown off by collisions escapes into space rather than falling back onto the surface. Or instead of being blasted out by the impact, the surface material is compressed.

The deep-pocket effect may be further enhanced by the Sun. When sunlight warms the dark material (probably organic dust) on the floors of the impact craters, it melts the ice and increases the depths of the crater floor. The surrounding ice evaporates and disperses.

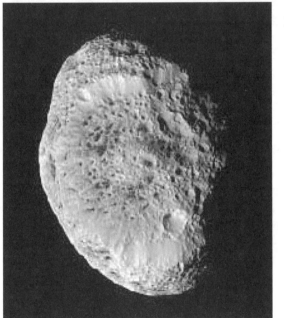

Figure 5

Iapetus

Iapetus, the third largest moon, is another weird satellite. It very strangely has one side that is extremely dark, while the other side is extremely bright (Figure 6). The dark side reflects less than 5% of the light from the Sun; the bright side reflects 50%. In other words, one side is 10 times brighter than the other side.

Figure 6

The satellite's dark side is the one that leads in its orbit. The prevailing theory was that Iapetus picks up dust sent into orbit by its companion moons. In 2009, images from NASA's Spitzer Space Telescope confirmed that idea, showing dust from another Saturnian satellite landing on Iapetus's leading side. Phoebe, orbiting in the opposite direction to Iapetus, appears to be the culprit.

Iapetus's slow rotation—80 days—also plays a role in the moon's unusual two-toned appearance. The Sun heats the dark areas enough to turn the ice into water vapor. However, the water vapor can freeze somewhere else, making the bright regions even more brilliant.

Both the bright and dark sides of Iapetus are covered with ancient impact craters. A concentration of dark material near the equatorial latitudes known as Cassini Reggio is probably an impact feature.

Iapetus has a higher density than the other Saturnian moons. It probably contains methane ice along with the water ice.

More (Mostly Miniscule) Moons

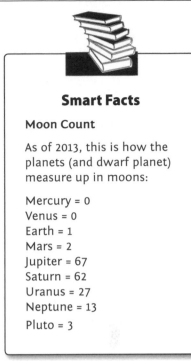

Discoveries made over the last decades have raised the tally of moons for Jupiter and Saturn both. Saturn has 62 moons (at last count), though only 13 with diameters of more than 50 kilometers. This count, of course, excludes the innumerable tiny ringmoons. In reality, ring particles and moons fall along a continuum so there is actually no specific cutoff diameter.

The best time to discover new moons is when Saturn's rings are on edge, when there is minimal glare.

Smart Facts

Moon Count

As of 2013, this is how the planets (and dwarf planet) measure up in moons:

Mercury = 0
Venus = 0
Earth = 1
Mars = 2
Jupiter = 67
Saturn = 62
Uranus = 27
Neptune = 13
Pluto = 3

Solar System Scoop

By the 20th century, all the names of the Titans had been assigned to discovered moons. The need for more names led to naming moons after various characters from Greek and Roman mythology or giants from other mythologies. With the exception of Phoebe, all the irregular moons are named after Inuit and Gallic gods and Norse giants.

Janus is one of the more interesting of the smaller moons (Figure 7a). Janus and Epimetheus share a single orbit and never stray more than 35 miles (50 kilometers) from each other. Roughly every four years, they switch places by exchanging momentum. The moons might once have been part of a parent body that was blasted apart in a collision.

Figure 7a

Janus's craters seem to be partly smoothed or erased by a layer of dust. Some of the other moons have this same feature. Figure 7b shows Janus in super-close-up, one of Cassini's closest views of the moon. Tiny Janus is less than 4% of Titan's diameter.

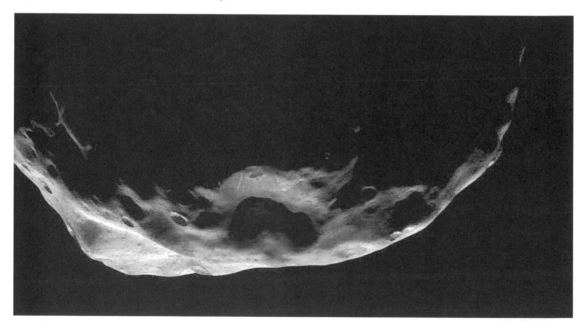

Figure 7b

Discovered in 1966, Janus was named for the Roman god of gates and doors, beginnings and endings, best known for being the two-faced god. (Janus might seem a more appropriate name for Iapetus.) The name Janus was originally proposed at the time of the moon's discovery, but it was not given the name until 1983 when Epimetheus was named.

PART FOUR

Cold and Distant

CHAPTER 17

Uranus: The Blue-Green Planet

In This Chapter

➤ William Herschel discovers Uranus

➤ Uranus's unusual color

➤ The dramatic discovery of the rings

➤ Voyager 2 flies by Uranus

➤ Twenty-seven moons, five major moons

Uranus was the first planet to be discovered through a telescope lens. Uranus is visible to the naked eye, but the dim, distant object was thought to be a faint star. That changed on March 15, 1781, when William Herschel, then a musician and amateur astronomer, peered through a telescope in his garden in Bath, England, and saw a slowly meandering object. The object had a disk, which would be an anomaly for a star. Herschel wrote in his journal that he witnessed a "nebulous star" or a comet. Two days later, the object had moved in the sky, convincing Hershel that what he saw was a comet. In fact, he presented his discovery to the Royal Society in a paper entitled "Of a Comet," intended for the society's publication, *Philosophical Transactions*.

From the onset, there was some speculation that the object was just as likely to be new planet as a comet. As its position was tracked by Herschel, astronomers at Cambridge, and astronomers outside of Great Britain intrigued by the new discovery, it became clear that the object moving among the stars was a distant planet, not a comet. Ironically, Herschel was more reluctant to promote his discovery to the status of planet than his peers were; perhaps he was just being prudent. By 1783, the object was universally recognized as a new planet, and Herschel was ready to take full credit as the discoverer of a "Primary Planet of Our Solar System." He immediately gained the fame he deserved.

Herschel wanted to name the planet in honor of his new patron, King George III, calling it *Georgium Sidus*, or "George's Star." The name was not very popular beyond Britain, and in any event, all the other planets were named after gods from Greco-Roman mythology. Uranus, after the god of the sky (Ouranos in Greek) was the popular choice.

Solar System Scoop

Throughout history, there has been an affinity between astronomy and music. William Herschel was originally a musician in the regimental band of Hanoverian Guards of Northern Germany. Following the disastrous battle of Hastenbeck of the Seven Years War, he left his regiment for England, where he became a music instructor, and in 1766, organist in the octagonal chapel at Bath (in 2010, the meeting place of the 7th Inspiration of Astronomical Phenomena).

Herschel's sister Caroline soon joined him. As we discussed in the history of modern astronomy, brother and sister both shared a love of astronomy. William became a master at building telescopes, constructing instruments superior to those made by professionals. His discoveries with those telescopes include Uranus, Saturn's moons Mimas and Enceladus, and Uranus's moons Titania and Oberon.

Herschel also created the General Catalogue of celestial objects, later the New General Catalogue, whose numbering scheme we still use today. Herschel's home at Bath where he discovered Uranus is now the Herschel Museum of Astronomy.

Solar System Scoop

Jupiter, Saturn, Uranus, and Neptune are often referred to as gas giants—fitting for colossal planets with no solid surface. But farther out in space, Uranus and Neptune are much colder and different enough from Saturn and Jupiter to be more accurately labeled "ice giants."

One reason Uranus eluded discovery as a planet (rather than a faint star) is its very slow revolution. One year on Uranus equals 84 Earth years.

Smart Facts

Uranus Facts

Uranus's orbit

Average distance from Sun	19.2 AU	2,875 million kilometers
Period by the stars	83.8 years	
Period from the Earth	370 days	
Orbit's eccentricity	0.05	
Orbit's tilt	0.8°	

Uranus the planet

Uranus's diameter	51,118 km	4.0 Earth's
Uranus's mass	14.5 times Earth's	
Uranus's density	1.3 times water	
Uranus's surface gravity	0.8 Earth's	
Sidereal rotation period	17 hours 14 minutes backward (on its side)	
Inclination of equator to orbit	98°	

Studying Uranus from Earth

At 20 A.U., Uranus is so far from us that telescopes based on Earth could see little more than a small disk. The tiny disk had a vaguely bluish-green color but no visible detail.

Uranus was known to have several small moons. Spectra of the ice giant's atmosphere revealed the presence of methane, which could account for the unusual color. Methane absorbs a lot of the orange and red from sunlight so that most of the light reflected back to us is green.

The clouds surrounding Uranus proved to be made of methane ice crystals, visible to us through a clear atmosphere of molecular hydrogen (H2). The blue-

Solar System Scoop

William Herschel initially described Uranus as a "nebulous star." Upon discovering a type of nebula looking rather disk-like and greenish like Uranus, he coined the term "planetary nebula." These objects are actually regions of gas ejected by stars like the Sun when they die; they have no relation to planets at all. But misnomer or not, the name planetary nebula remains.

green color results from an admixture of methane and the hydrogen molecules. Chemical reactions are extremely slow on cold, icy Uranus. Infrared observations show a temperature of -355 degrees F (-215 degrees C).

All of the planets have some odd characteristics. Uranus has a weird axial tilt. It spins on its axis at an angle so exaggerated that it is tipped on its side. The moons orbit the planet's equator, positioning them in planes almost perpendicular to the paths of the planetary orbits. A massive collision back in the chaotic early universe seems to have turned the giant planet halfway around.

Well into the twentieth century, Herschel's discovery was still notoriously difficult to study from Earth.

Occultations and an Astounding Discovery

A "eureka" moment finally came in 1977. James Elliot (then of Cornell University, now MIT) was flying over the South Pacific in an attempt to observe the atmosphere of Uranus from NASA's Kuiper Airborne Observatory. He was planning to measure the light from a background star as Uranus occulted it (moved in front of it and hid it). Due to the plane's moving location, the timing of the occultation was a bit uncertain. To compensate, Elliot began his observations more than half an hour before the time of ingress and kept them up more than half an hour after the calculated time of egress (that is, when Uranus began blocking the star and, minutes later, began uncovering it).

The scientist had expected to see the light from the star disappear gradually, in a minute or so, as it was bent and attenuated by Uranus's atmosphere. Only even before Uranus began moving in front of the star, and again when it was over, the detector picked up a series of dips in the starlight. Something else seemed to be coming in front of the star.

Elliot first reported that he picked up a group of moons surrounding Uranus. But he quickly realized that the dips before the main occultation and the dips after it were symmetrical. Most of our solar system planets have moons. Uranus was surrounded by circular rings! That discovery was nothing short of amazing.

Needless to say, these occultations occur very far from Earth's atmosphere. As a result, the precision is not degraded by the normal blurring caused by our atmosphere (in astronomical jargon, the "seeing"). All that's important is the relative speed of the occulting object and the star as they rise and set. It's an amazing technique for capturing distant objects. We'll get to this powerful technique again when we cover Pluto and the objects beyond it in the outer reaches of the solar system.

As marvelous as it was to discover Uranus has rings, they are nothing like Saturn's spectacular rings. The rings circling Uranus—11 to date—are very thin, in some cases no more than a few miles wide. At least some of the rings are held in position by shepherding satellites, the same phenomenon we see in the ringlets and narrower rings surrounding Saturn. The rings are extremely dark; they barely reflect 2% of the incoming sunlight.

The moons shepherding the rings are even puny for moonlets. They lack the mass to hold the rings' particles in place for more than roughly 100 million years. That means we are probably seeing new moons formed from the tiny moons breaking apart.

Voyager 2 Heads Toward Uranus

After harnessing Jupiter's gravity to reach Saturn, Voyager 2 got a boost from Saturn's gravity and headed farther out toward Uranus. It arrived at its closest point to Uranus on January 24, 1986, coming within 50,600 miles (81,500 kilometers) of the ice giant's cloud tops.

In its short time near the Uranus system, Voyager 2 transformed objects that had appeared as blurs or specks into celestial objects with distinct characteristics. Uranus turned out to have even more rings and moons. The spacecraft revealed the additional rings and examined the already known rings from a backlit perspective. The image showed that the dust in Uranus's rings is relatively sparse compared to the dust content in the rings of Saturn and Jupiter. Cordelia, Ophelia, Bianca, Cressida, Desdemona, Juliet, Portia, Rosalind, Belinda, Perdita, and Puck joined Uranus's retinue of moons. The known moons each gained their own personalities.

Even close up, Uranus's icy, gaseous surface was bland. Any features in the planet's atmosphere were obscured by a high-level haze (Figure 1). Even the highest contrast computer analysis showed nothing more than a few elongated clouds. These clouds are probably methane ice lifted up high by convection in Uranus's atmosphere.

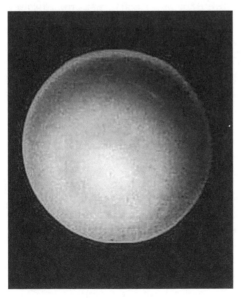

Figure 1

Inside the Blue-Green Planet

The composition of Uranus compared to Jupiter and Saturn suggests they evolved somewhat differently. Uranus has the second lowest density of any planet in our solar system, next to Saturn. Unlike Saturn and Jupiter, Uranus is carbon-rich. It is thought to contain 20 times more carbon than our Sun.

Voyager 2 detected a powerful magnetic field. In fact, Uranus has a magnetic field 50 times stronger than Earth's. The discovery of the magnetic field per se was not unexpected. What came as a shock was its peculiar angle. Uranus's magnetic field is tilted two-thirds of the way over, compared with the planet's axis of rotation. And for another odd characteristic, the magnetic field is not at the planet's center, but off-center by one-third of the radius of Uranus.

As a result of the magnetic field, Uranus has a magnetosphere with belts of charged particles. These belts are akin to the Van Allen belts surrounding our Earth.

Radio observations picked up bursts of radiation every 17.5 hours. They seem to emanate from deep under the surface. This may turn out to be the true rotation period of Uranus. It's a bit slower than the atmospheric rotation deduced from wavelength shifts and cloud movements.

Adaptive Optics on Earth, Hubble Telescope in the Sky

Computers and telescopes make powerful allies in the quest for learning as much as possible about the universe from Earth. Electronic imaging systems like computer chips giving off electrons from each pixel when hit by light provide optimum sensitivity. By combining the two technologies, scientists have developed a technique called adaptive optics that compensates for the blurring caused by Earth's atmosphere. Images produced by adaptive optics are far superior to traditional telescope images.

Using this technique, astronomers like Heidi Hammel and her colleagues at the Space Science Institute have been studying Uranus with terrestrial telescopes. They've discovered clouds and are able to track their circulation. Thanks to their observations, we are learning more about the effects of seasonal changes as each of the poles (as well as the middle latitudes) emerges from the darkness into the sunlight during Uranus's very long year. Herschel would probably be amazed to know that scientists on Earth have been able to follow the distant Uranus for much of its 84-Earth yearlong journey around the Sun.

While astronomers using adaptive optics study Uranus from Earth, the Hubble Space Telescope joins them in providing us with high-quality images (Figure 2). Images from adaptive optics and the Hubble cameras rival each other in high resolution.

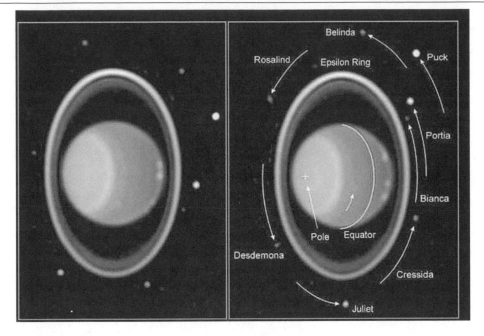

Figure 2

Moons of Uranus

Voyager 2 conducted the most thorough investigation of Uranus's moons. Going outward from Uranus, the main moons are Miranda, Ariel, Umbriel, Titania, and Oberon (Figure 3). If the names sound familiar, all the moons are named for characters from the works of William Shakespeare and Alexander Pope. The current tally is 27.

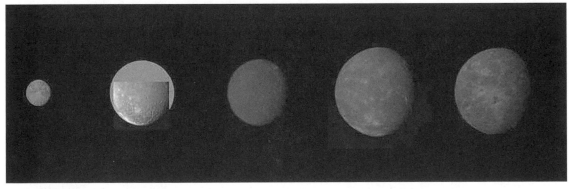

Figure 3

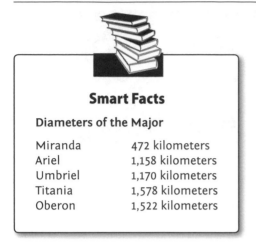

Voyager 2 flew close enough to Miranda to measure its gravitational pull. The masses of the other moons were deduced from their gravitational effects on each other. The densities of the moons put them in the same class as the moons of Jupiter and Saturn; that is, mainly rock and ice. The ice is composed of water, ammonia, methane, and other chemicals.

Smart Facts

Diameters of the Major

Miranda	472 kilometers
Ariel	1,158 kilometers
Umbriel	1,170 kilometers
Titania	1,578 kilometers
Oberon	1,522 kilometers

Solar System Scoop

Brush Up Your Shakespeare (and Pope)

John Herschel, William's son, named Oberon and Titania for the King and Queen of the Fairies in Shakespeare's *Midsummer Night's Dream*. His father discovered the moons in 1787. Umbriel and Ariel were discovered by William Lassell in 1851 and were given names from Pope's *The Rape of the Lock*. Small Miranda was discovered in 1948 by Gerard Kuiper and named for the main character in Shakespeare's *The Tempest*. The moons discovered by Voyager 2 are also named for characters in Shakespeare and Pope.

Ariel

Like most moons, Ariel is covered with craters. The craters span about three to six miles (five to 10 kilometers) across. Ariel's terrain shows geological faults and scarps crossing successions of craters. These features were probably formed from an expansion that stretched the moon's crust.

There are also less cratered regions on Ariel's landscape, where valleys and other areas were partly filled with younger material. These younger deposits have covered the faults crossing the valleys. Twisting trenches formed at some later point, which could imply that liquid once flowed over the icy moon.

Water could never melt on frigid Ariel, but it is possible that eons ago there were tidal forces generating enough heat for water to flow. Or there might have been a mixture of ammonia and water flowing over the terrain.

Figure 4. The most detailed Voyager 2 photograph of Ariel.

Umbriel

Umbriel is heavily cratered and very dark. With an albedo of only 16%, it is much less reflective than its companion moons. Covered all over in craters, it seems to have had little geological activity, definitely less than any of the other moons. Umbriel's surface resembles the terrain of our own Moon's highlands.

Umbriel's ancient surface may date back to the dawn of the solar system 4.5 billion years ago. All that's altered its landscape are the craters from so many collisions.

Umbriel has one curious feature: a bright 90-mile (140-kilometer) ring near its equator. It could be frost on or surrounding a deep crater. It had to have happened after Umbriel's surface became so dark. None of Uranus's other moons are so dark, meaning that whatever caused the darkening effect came from the moon itself. A major collision or an eruption could be the source.

Unlike most moons, Umbriel has no bright, rayed craters. The question of why may be answered someday; for now, it remains a mystery.

Figure 5. A bright ring is visible near the north pole in this photograph of Umbriel.

Titania

Titania is dotted with circular depressions that are probably impact craters (Figure 7). Bright rays emanating from some of the craters indicate that they come from recent impacts. A fault canyon traverses a sizable portion of Titania's surface. Crossing at various angles, the gorges probably formed when the moon's crust expanded: signs of a geologically active interior.

Titania has both young and old terrain. The more recent deposits cover the biggest craters, which were probably formed in the earliest years of the solar system.

Figure 6. Titania, photographed from Voyager 2.

Oberon

Oberon is Uranus's largest moon. This image of Oberon taken by Voyager 2 (Figure 8) shows bright patches, probably matter dug up and ejected by impacts.

Oberon is marked by several expansive, dark, circular regions. The dark material may be similar to the dust deposited on Iapetus's trailing side. In Oberon's case, however, the cause is probably dark material that escaped from the moon's interior eons ago as opposed to errant dust from a companion moon.

Figure 7. There are high albedo patches on Oberon that match rays thrown out from impacts.

Miranda

The smallest of Uranus's major moons, Miranda was not discovered until the mid-twentieth century. It turned out to have the most interesting surface of all the moons. It looks like an object that was torn apart and the pieces thrown back together in some very haphazard way (Figure 9). Though Miranda has no discernible tidal heating at the present time, at some point in the remote past, the moons might have had different orbits, and Miranda's interior might have been hot enough to weld the pieces together.

There is evidence that Miranda once had internal activity. Near its south pole, there is a trapezoidal region (roughly 200 kilometers on a side) with sharp corners and edges on its ridges and outer boundary. Nothing other than geological activity on Miranda could have caused those sharp cuts.

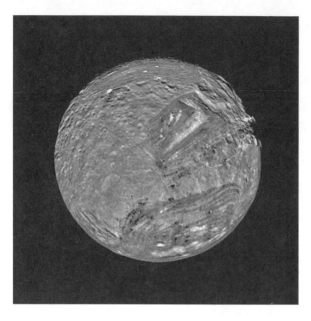

Figure 8. This image of Miranda is a composite comprised of several distant views.

Satellite orbits

Satellite	Orbital size (km)	Eccentricity	Inclination (°)	Period (days)
Ariel	190900.	0.0012	0.041	2.520
Umbriel	266000.	0.0039	0.128	4.144
Titania	436300.	0.0011	0.079	8.706
Oberon	583500.	0.0014	0.068	13.46
Miranda	129900.	0.0013	4.338	1.413
Cordelia	49800.	0.0003	0.085	0.335
Ophelia	53800.	0.0099	0.104	0.376
Bianca	59200.	0.0009	0.193	0.435
Cressida	61800.	0.0004	0.006	0.464
Desdemona	62700.	0.0001	0.113	0.474
Juliet	64400.	0.0007	0.065	0.493
Portia	66100.	0.0001	0.059	0.513
Rosalind	69900.	0.0001	0.279	0.558
Belinda	75300.	0.0001	0.031	0.624
Puck	86000.	0.0001	0.319	0.762
Perdita	76417.	0.0116	0.470	0.638
Mab	97736.	0.0025	0.134	0.923
Cupid	74392.	0.0013	0.099	0.613
Caliban	7231100.	0.1812	141.529	579.73
Sycorax	12179400.	0.5219	159.420	1288.38
Prospero	16276800.	0.4445	151.830	1978.37
Setebos	17420400.	0.5908	158.235	2225.08
Stephano	8007400.	0.2248	143.819	677.47
Trinculo	8505200.	0.2194	166.971	749.40
Francisco	4282900.	0.1324	147.250	267.09
Margaret	14146700.	0.6772	57.367	1661.00
Ferdinand	20430000.	0.3993	169.793	2790.03

CHAPTER 18

 # Neptune: Outer Ice Giant

In This Chapter

➤ First planet predicted before its discovery

➤ Neptune's weird magnetic field

➤ Neptune's clumpy rings

➤ Discovering Neptune's moons

➤ Triton, Neptune's main moon

Icy Neptune, the densest of the gas giants, is the farthest planet from the Sun. At 30 AU, any sunlight reaching the planet is very weak.

Neptune has an interesting history. It was the first planet predicted by mathematical calculation rather than found by direct observation. During the 1840s, British university student John Couch Adams and French scientist Urbain Leverrier each independently studied Uranus's orbit and realized there was something odd in its shape. According to Kepler's first law of planetary orbits, a planet should have an elliptical orbit if it is only influenced by the Sun. Something was disrupting Uranus's presumed elliptical path. Adams and Leverrier each figured out the position of the yet unknown planet that was throwing off Uranus's orbit.

In the end, it was the established scientist rather than the young student who successfully gained support in finding the predicted planet. Leverrier sent a letter to astronomer Johann Galle of the Berlin Observatory, urging him to turn their telescopes skyward based on his estimates of the new planet. On the night of September 23, 1846, the same day Galle received Leverrier's letter, he observed a "star" at the predicted position. There was no record of any such star in the most up-to-date star map.

Neptune's discovery spurred a fight between the British and French over who should get credit. Over time, an international consensus decided that Leverrier and Adams should receive equal credit, but history has since favored Leverrier for predicting the planet and more important, persuading fellow astronomers to search for it.

In an interesting twist, Galileo saw Neptune, thinking it was a star. Notes and sketches of Jupiter and its moons from December 1612 include notations about a "star" that was very close to Jupiter on December 28th. He observed the object again on January 27, 1613. Its position matches the predicted position of Neptune.

Twentieth century telescopes could make out some features on Neptune's disk, though very indistinct. However vague the images were, Neptune's surface promised to be more interesting than bland Uranus. After its sojourn at Uranus, Voyager 2 continued out into the solar system and arrived at Neptune in 1989.

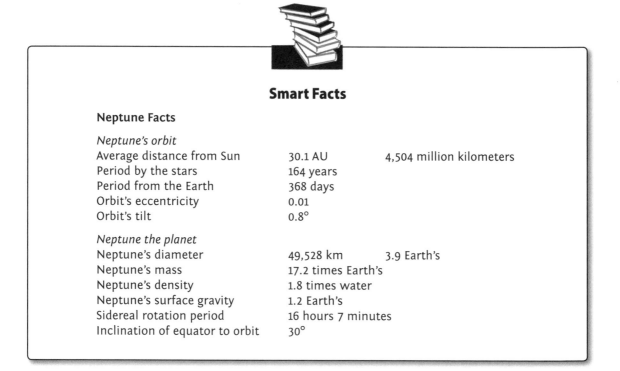

Smart Facts

Neptune Facts

Neptune's orbit

Average distance from Sun	30.1 AU	4,504 million kilometers
Period by the stars	164 years	
Period from the Earth	368 days	
Orbit's eccentricity	0.01	
Orbit's tilt	0.8°	

Neptune the planet

Neptune's diameter	49,528 km	3.9 Earth's
Neptune's mass	17.2 times Earth's	
Neptune's density	1.8 times water	
Neptune's surface gravity	1.2 Earth's	
Sidereal rotation period	16 hours 7 minutes	
Inclination of equator to orbit	30°	

Bode's Law

In 1778, the German astronomer Johann Bode popularized a rule that seemed to govern the spacing of the planets. Sometimes called the Titius-Bode law, after Johann Titius, who developed the rule 12 years earlier, its popularity stretched into the nineteenth century.

According to Bode's law (or rule):

After 0, begin a sequence with 3, and then keep doubling the last number:

0	3	6	12	24	48	96	192	384

Then add 4 to each number:

4	7	10	16	28	52	100	196	388

And then divide by 10:

0.4	0.7	1.0	1.6	2.8	5.2	10.0	19.6	38.8

These numbers seem to do a good job of matching the spacing of the planetary orbits in astronomical units:

Mercury	Venus	Earth	Mars	Jupiter	Saturn	Uranus	Neptune
0.39	0.72	1.0	1.52	5.2	9.5	19.2	30.1

The discovery of Uranus was consistent with Bode's prediction. The gap between Mars and Jupiter can be explained by the asteroid belt, so that also works with Bode's system. However, the match between Bode's numbers and the objects' positions in space ends with Uranus. Once when we get to Neptune, the rule loses its credibility.

Perhaps the most interesting thing about Bode's law is that it worked so well…up to a point. It's actually nothing more than numerological coincidence. But it's not difficult to see why it would have intuitive appeal.

The Great Dark Spot

After glimpsing Neptune through terrestrial telescopes, astronomers eagerly awaited the high-resolution images from Voyager 2. The first exciting images revealed a Great Dark Spot that was similar in shape to Jupiter's Great Red Spot, but even bigger compared to the relative size of the planets (Figure 1).

By putting together a series of photographs taken at the same phase of Neptune's rotation and showing the Great Dark Spot, scientists could track its rotation. The Great Dark Spot turned out to share many characteristics with the Great Red Spot. Like the Great Red Spot, the Great Dark Spot rotates in a counterclockwise direction. Since both are located in the southern hemisphere of their parent planets, the Great Dark Spot is also anti-cyclonic. And because of that, it is also located in a high-pressure region.

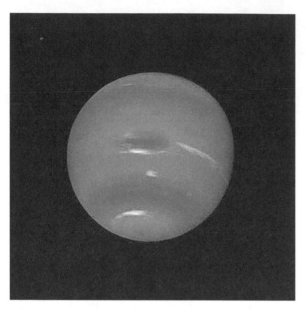

Figure 1

Voyager 2 observed clouds of methane ice form at the edge of the Great Dark Spot. The methane-rich gas is forced upward by the spot's high pressure. The clouds looked similar to the cirrus clouds we see in the sky. Voyager 2 captured images of various clouds in Neptune's atmosphere. What it did not uncover in its short stay at Neptune is that the Great Dark Spot was not as much like the Great Red Spot as scientists thought.

Adaptive Optics and Hubble Turn to Neptune

Adaptive optics, as we learned, allow scientists to hone in and see sharp, detailed images of remote objects from their terrestrial base. Adaptive optics make images of Neptune distinct enough to show bands of clouds, though they are only visible in the infrared. (The longest wavelengths are the easiest to make. Adaptive-optics systems are continually improving, but the techniques don't work as of yet in the visible range.)

The trusty Hubble Space Telescope has also been turned to Neptune. And Hubble provided scientists with an unexpected surprise: the Great Dark Spot was gone! It was just a few years after Voyager 2 discovered the Great Dark Spot, but it was no more (Figure 2). As similar as the two phenomena seemed to be, the Great Dark Spot was not a storm raging for hundreds and hundreds of years like the Great Red Spot.

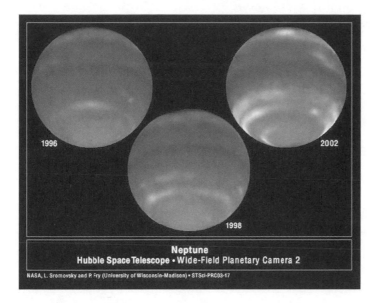

Figure 2

With an orbit around the Sun equivalent to 164 Earth years, seasons change *very* slowly on Neptune. Ongoing observations with the Hubble Space Telescopes and adaptive optics from mountaintop observatories show us how Neptune's weather evolves.

Inside the Ice Giant

Neptune's average temperature is -353 degrees F (-214 degrees C). Cold as that is, it's roughly the same as Uranus's temperature, even though Neptune is farther out from the Sun. The reason for this is that Neptune radiates about 2.7 times more energy than it receives from the Sun. In other words, Neptune has some internal heating mechanism, which Uranus apparently lacks.

Neptune has the highest density of the gas giants, implying the presence of heavy elements. It might have a small core with gravity too weak to attract an immense atmosphere of hydrogen and helium, even as cold as it is. Alternately, there might have been less hydrogen and helium that far out in the solar system when Neptune formed. It's also possible that the solar wind blew away a lot of the lighter elements before Neptune evolved.

Like Uranus, Neptune gives off radio bursts that were picked up by Voyager. Neptune's bursts happen every 16.11 hours, most probably the speed of rotation of its interior. Both Neptune and Uranus have interiors that rotate faster than the surface winds at their equators. No other planets with atmospheres have that property. This distinction is especially valuable in the context of comparative planetology. It might give us a better understanding of our own winds and weather.

Neptune's Very Odd Magnetic Field

Neptune's magnetic field is even stranger and stronger than its neighboring ice giant's powerful, tilted magnetic field. After the initial surprise of Uranus's magnetic field, scientists were aware that a magnetic field could be strong and off-center. They were still astounded when Neptune turned out to have a magnetic field that surpassed Uranus in both those effects.

In addition to being very powerful, Neptune's magnetic field is tipped halfway over (47 degrees) and offset by 55 degrees. In other words, its center is more than halfway out from the planet's center.

Since Uranus is tipped so far on its side, some scientists figured that the tipped magnetic field could have been caused by the same collision that threw it over. However, that explanation could not apply to Neptune, which rotates basically upright like a typical planet. Neptune's off-kilter magnetic field called into question the collision theory for Uranus's magnetic field.

The current theory is that the magnetic fields on both ice giants are formed in a shell of electrically conductive liquid. These shells lie outside the cores of the planets, in contrast to Earth or Jupiter, whose magnetic fields are thought to emerge from inside the cores.

Neptune's Clumpy Rings

Since all the other gas giants have rings, it seemed probable that Neptune would have rings too. Given the discovery of Uranus's rings as Uranus occulted a star, a number of expeditions set out to try to detect Neptune's rings in the same way.

Several such expeditions were conducted in 1983. Curiously, some scientists on Mauna Kea detected an occultation, but others at nearby telescopes did not. The driving question was why? It could be that Neptune had "ring arcs," or partial rings, where the material fails to spread out enough to fill the whole ring. In that case, a star would be blocked by the filled part of the ring, while the starlight would shine through the clear, unfilled portion.

The answer would have to wait for Voyager 2's arrival in 1989. The Voyager images showed rings that were narrow but very clumpy (Figure 3). Some parts of Neptune's rings contain a hundred times more small, dusty particles than the rings of Uranus and Saturn. At the same time, particles in many parts of Neptune's rings are very sparse. The thick clumps are probably areas where ringmoons continually collide and smash each other to keep replenishing the dust in the rings.

Figure 3

The two main rings are named for Adams and Leverrier, whose calculations predicted the presence of Neptune. A weaker ring is named for Galle, the astronomer who first observed the new planet. The Adams ring has clumps named after the cry of the French Revolution: Liberté, Egalité, Fraternité. (Interesting after the fight between Britain and France over credit for the new planet that the slogan should be assigned to the Adams ring rather than the Leverrier ring.)

Astronomers observing the rings have seen some changes since the first images two decades ago. One of the ring arcs has faded significantly. At the rate it appears to be happening, it will disappear completely within the next hundred years.

Moons of Neptune

Neptune's largest moon Triton was discovered in 1846 by William Lassell, a mere seven days after the discovery of the planet itself. Big and barely fainter than Neptune, it was not difficult to see once telescopes were aimed at the new planet. It was almost 100 years until another moon was discovered. Nereid, the second moon, is more than 250 times fainter than Triton.

The next group of moons was not

Solar System Scoop

Neptune (Poseidon in Greek) is the Roman god of the sea, brother of Jupiter and Pluto. Interestingly, as an alternative to "George's Star," there was a move to name William Herschel's discovery Neptune in honor of the strong British Navy. That time Neptune lost out to Uranus, but it became the name of the next new planet.

Triton was a sea messenger and a son of Poseidon. The Nereids were sea nymphs in Greek mythology.

discovered until Voyager 2 reached the Neptune system. The spacecraft found six additional moons, and five more were discovered by astronomers at Mauna Kea in 2002 and 2003. These moons are extremely faint; even Nereid outshines them by a factor of 250 or more.

Triton: The Biggest and Brightest Neptunian Moon

Neptune's largest satellite is actually rather puny when compared to a giant like Titan. Even our own moon is roughly 20% bigger in diameter than Triton. For the Neptune system, Triton is the only satellite massive enough to be round.

Triton has a retrograde orbit with respect to its parent planet's rotation. Though many small, distant, irregularly shaped moons have retrograde orbits, Triton's backward orbit is unique among the large inner moons in our solar system. Orbiting very close to Neptune, and in almost a perfect circle, Triton is probably an asteroid captured by the massive ice giant. Locked into orbit by Neptune's gravity, the same side of the moon always faces the planet.

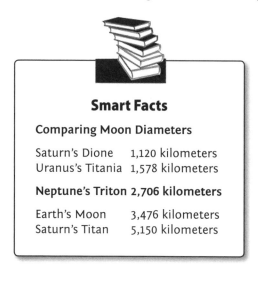

Smart Facts

Comparing Moon Diameters

Saturn's Dione   1,120 kilometers
Uranus's Titania   1,578 kilometers

Neptune's Triton 2,706 kilometers

Earth's Moon   3,476 kilometers
Saturn's Titan   5,150 kilometers

Beyond being the biggest, most massive Neptunian satellite, Triton has the highest density of any other moon attached to a gas giant, with the exception of Jupiter's Io and Europa. Triton's density, 2.1 times as dense as water, implies a composition of rock and water ice, with a ratio of rocky material to ice of 2:1.

Triton's atmosphere had been discerned by Earth-based spectroscopic recordings even before Voyager 2 reached the Neptune system. It was then examined in detail through occultations. Intriguing scientists even more, its atmosphere turned out to be made up primarily of molecular nitrogen, a property shared by our own planet and Titan.

Triton was Voyager 2's last destination. That was actually advantageous, because the spacecraft could fly very close to the moon without jeopardizing any future activity. Scientists were eager to learn more about the Neptunian moon. At the time, though, they had no idea whether its atmosphere would be clear enough for Voyager's cameras to capture shots of its surface.

Fortunately, Triton's atmosphere was nothing like smoggy Titan's. It turned out to be very clear, and Voyager sent back excellent quality, exciting images of its surface (Figure 4). The photographs of Triton's southern regions are especially interesting, revealing among other things, a polar cap with reddish-tinged ice. Scientists are always intrigued by a reddish or orangey hue because it is typically caused by organic material—organic implying that life might have arisen there at some point.

It was late spring on Neptune when Voyager snapped it in 1989, and photographs show the edges of Triton's polar ice caps receding. Given Neptune's extraordinarily long year, it is still late spring in the Neptune system.

Figure 4

Any organic material on Triton probably formed from a combination of ultraviolet light from the Sun and particles from Neptune's magnetosphere hitting the methane in Triton's surface and atmosphere. Triton's surface is composed mainly of nitrogen frost, with traces of condensed methane, carbon dioxide, and carbon monoxide.

Triton is home to an expansive region that resembles the skin of a cantaloupe melon. This "cantaloupe terrain" is composed of an odd succession of grooves and depressions on a scale of roughly 20 miles (30 kilometers). Though there are few impact craters, the region is thought to be Triton's most ancient landscape.

Voyager images revealed cliffs and craters too high and too clear to be dominated by pure methane ice. This phenomenon implies that water ice plays an important role in the structures within this region.

For a moon, Titan has relatively few impact craters. Most are on its leading hemisphere, with more exposure to violent collisions. The culprits are most likely comets, since there are not many asteroids to collide with that far out in the solar system.

A particularly awesome feature revealed by Voyager close-ups was a series of roughly 50 parallel dark streaks (Figure 5). Darkened methane ice heated up by the Sun probably vaporized nitrogen ice lying just beneath Triton's surface. The nitrogen then escapes through vents on the surface, blowing the dark material upward. The parallel streaks imply that Triton's atmosphere has a steadily blowing wind. Under the dark streaks, the ice changes with Neptune's (very long) seasons. Thus, the streaks are extremely young astronomically; no more than about 100 years old.

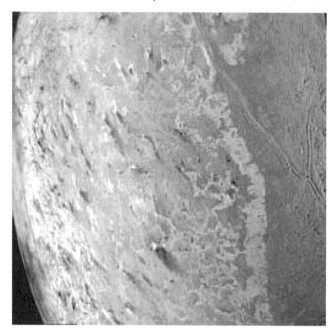

Figure 5

As a captured asteroid, Triton would have been subject to powerful tidal forces. As a result, tremendous energy in the moon's interior would melt it and keep it molten. Calculations suggest that Triton has a core making up a substantial portion of its interior.

Triton's landscape proved to be at least as interesting as scientists anticipated. At the same time, Voyager 2 imaged only 40% of its surface. Its unexplored surface might have more impact craters, or perhaps something even more unusual than parallel dark streaks or cantaloupe terrain.

Global Warming on Icy Triton?

Triton's atmosphere makes it an excellent candidate for observing through occultations. By distorting and attenuating the starlight, the atmosphere prolongs the time that the star disappears from view and then reappears. The catch is that Triton only passes in front of a bright-enough star once every few years.

In 1998, the Hubble Space Telescope was perfectly positioned to catch Triton occulting a star six times brighter than the Neptunian moon. As Triton blocked out the star, there was a steep drop in the total light. Detailed analysis revealed that Triton was slightly warmer than the temperature recorded by Voyager eight years earlier. It does have an atmosphere similar in composition to Earth and Titan—even out in the nether reaches of our solar system. However, it is much, much thinner, with a pressure less than 1/70,000 that of Earth's at the surface.

Satellites of Neptune

Satellite	*Mean radius* (km)	*Mean density* (g/cm³)	*Geometric Albedo*
Triton	1352.6 ± 2.4	2.064 ± 0.011	0.756 ± 0.041
Nereid	170. ± 25.	1.5	0.155
Naiad	33. ± 3.	1.3	0.072
Thalassa	41. ± 3.	1.3	0.091
Despina	75. ± 3.	1.3	0.090
Galatea	88. ± 4.	1.3	0.079
Larissa	97. ± 3.	1.3	0.091
Proteus	210. ± 7.	1.3	0.096
Halimede	31.0	1.5	0.04
Psamathe	20.0	1.5	0.04
Sao	22.0	1.5	0.04
Laomedeia	21.0	1.5	0.04
Neso	30.0	1.5	0.04

Satellite orbits

Satellite	Orbital radius (km)	Eccentricity	Inclination (deg)	Period (days)
Triton	354759.	0.0000	156.865	5.877
Nereid	5513818.	0.7507	7.090	360.13

Satellite	Orbital radius (km)	Eccentricity	Inclination (°)	Period (days)
Naiad	48227.	0.0003	4.691	0.294
Thalassa	50074.	0.0002	0.135	0.311
Despina	52526.	0.0002	0.068	0.335
Galatea	61953.	0.0001	0.034	0.429
Larissa	73548.	0.0014	0.205	0.555
Proteus	117646.	0.0005	0.075	1.122

Satellite	Orbital radius (km)	Eccentricity	Inclination (°)	Period (days)
Halimede	16611000.	0.2646	112.712	1879.08
Psamathe	48096000.	0.3809	126.312	9074.30
Sao	22228000.	0.1365	53.483	2912.72
Laomedeia	23567000.	0.3969	37.874	3171.33
Neso	49285000.	0.5714	136.439	9740.73

CHAPTER 19

Pluto and Charon: From Planet to Dwarf Planet

In This Chapter

> ➤ Clyde Tombaugh discovers Pluto
> ➤ Pluto's moon Charon discovered
> ➤ Faint, tiny moons Nix and Hydra
> ➤ Pluto, always peculiar for a planet
> ➤ New Horizons' long journey to Pluto

From the time it was discovered in 1930 to 2006, Pluto, the small rocky object orbiting the Sun in a weirdly eccentric path in the outer solar system was classified as a planet. That year, it was demoted to "dwarf planet" (and not without a big fight).

There had been nine planets in our solar system for so long that most of us were complacent about the number. Historically, the number of planets has changed many times. Up until 1781, when William Herschel discovered Uranus, there were six planets: our own planet Earth, and the "wanderers" Mercury, Venus, Mars, Jupiter, and Saturn, known to the ancients. At different times, the Sun and Moon were regarded as planets. In early days, Earth was not. Uranus brought up the count to seven planets, and Neptune brought the number to eight.

There were other variations in the planet count. The first asteroids were discovered in 1801 and that first group became part of the planet tally, bringing the number to about 50. Then the asteroids were dropped from the count, and the number went back to eight.

As we know from all the collisions, our solar system is crowded with objects, and there are ongoing discoveries. New discoveries in the outer reaches of the solar system helped to seal Pluto's fate.

Pluto is actually one among hundreds of objects orbiting beyond the ice giants. Many of them have the size, mass, and shape to be considered dwarf planets. If Pluto's demotion came as a shock, the controversy led to a new classification of "dwarf planet" and to defining a "planet" for the first time ever. Improbably, the term "planet" had never been formally defined.

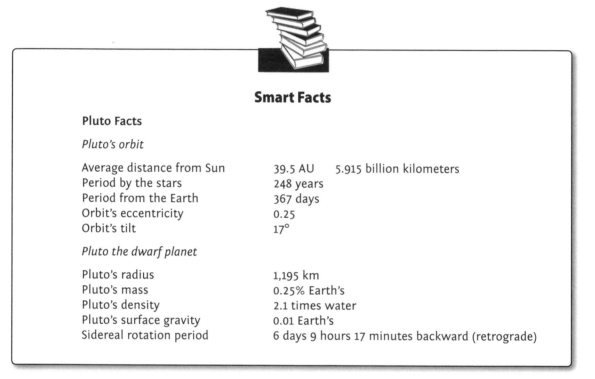

Smart Facts

Pluto Facts

Pluto's orbit

Average distance from Sun	39.5 AU	5.915 billion kilometers
Period by the stars	248 years	
Period from the Earth	367 days	
Orbit's eccentricity	0.25	
Orbit's tilt	17°	

Pluto the dwarf planet

Pluto's radius	1,195 km
Pluto's mass	0.25% Earth's
Pluto's density	2.1 times water
Pluto's surface gravity	0.01 Earth's
Sidereal rotation period	6 days 9 hours 17 minutes backward (retrograde)

Discovering Pluto

The Lowell Observatory in Flagstaff, Arizona, is a legacy of Percival Lowell, who founded it in 1894. The wealthy Bostonian was fascinated by astronomy. His special interest was Mars, and he wrote several books about the red planet and its alleged canals, even imagining what life on Mars would be like. But Lowell also had another obsession—he was convinced that there was another planet beyond Neptune.

Even with the knowledge of Neptune's gravitational effect on Uranus's orbit, there were still a few blips in its orbit that could not be explained by Neptune. The presumed planet was called "Planet X," and mathematical calculations predicted where to search for it in the sky.

When Lowell died in 1916, the observatory carried on with his exploration of Mars—and his search for Planet X. Clyde Tombaugh, a talented young amateur astronomer, was chosen to conduct the search. The meticulous Tombaugh took myriad images of the sky and scrutinized hundreds of thousands of photographs of individual stars. He methodically compared pairs of plates taken two weeks apart to determine if any objects in them had changed positions. A device called a blink comparator allowed him to quickly switch back and forth between the plates and discern any apparent motion.

On February 18, 1930, Tombaugh had been searching for almost a year. Examining photographic plates taken on January 23rd and January 29th of that year, he saw that a faint object had moved. An image from January 21st, though poor in quality, confirmed that Tombaugh had indeed found the presumed planet. He brought his discovery to Lowell's director, telling him simply, "I've found your Planet X." After additional confirmation, news of the discovery was sent to the Harvard College Observatory on March 13, 1930. Soon the world knew that the solar system had a ninth planet.

Solar System Scoop

The name Pluto was the idea of an 11-year-old English girl named Venetia Burney. Venetia thought the name of the god of the underworld was especially well-suited for a place that was presumably cold and dark. Her grandfather sent in the suggestion and the name became official on March 24, 1930.

Notably, the first two letters of Pluto are the initials of Percival Lowell, and Pluto's astronomical symbol is created from the letters "PL." No doubt the Lowell family had this in mind when they approved the young girl's suggestion.

The new planet may have also inspired Walt Disney's name for a new cartoon character. As we discussed earlier, Disney had tremendous respect for scientific discovery. Mickey Mouse's canine friend Pluto made his debut in 1930.

Venetia Burney died in 2009. A wonderful short video was made about her and the discovery of Pluto.

Imaging Pluto

The best images of Pluto come to us from the Hubble Space Telescope (Figure 1). The object is so small and distant that even Hubble images lack full resolution detail. On Earth, only telescopes using adaptive optics can successfully image Pluto.

Figure 1 portrays Pluto's disk in a pair of small insets, courtesy of the Hubble Space Telescope's Advanced Camera for Surveys. The global maps are created by computer-image processing based on Hubble photographs. The pixels in the original images are more than 100 miles (160 kilometers) across, so no smaller details are visible.

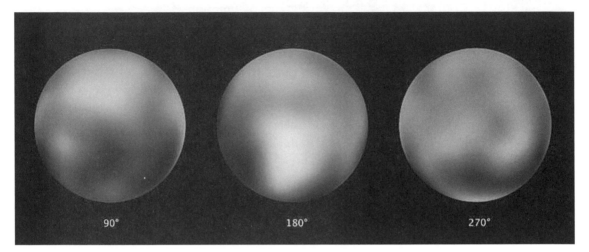

Figure 1

From what we can see, Pluto seems to have roughly a dozen major regions. However, with details obscured there is no way of knowing what causes the various light and dark shadings. They could come from differences in terrain such as craters or basins. Or they might be related to frost. Frost most likely accounts for the brightness at the northern polar cap. The area around the equator is dark.

Charon and Two Faint Moons

In 1978, James Christie of the U.S. Naval Observatory noticed that Pluto sometimes appeared with a bulge. It was especially noticeable when the images were very clear and sharp (when the seeing was good, as astronomers say). What's more, the bulge seemed to change its position and its movements matched Pluto's 6.4-day rotation period.

Christie had discovered a moon. He named it Charon, after the boatman of Greek mythology who ferried the dead across the river Styx to Pluto's underworld kingdom.

As it happened, the discovery of Charon began Pluto's downfall as a full planet. The moon's presence allowed Pluto's mass to be accurately gauged for the first time. It turned out to be only 1/400th the mass of the Earth—way smaller than any previous estimate.

When Tombaugh went searching for Planet X, the assumption had been that the deviations in Uranus's orbit were caused by an object roughly as massive as our own planet. Puny Pluto could not have possibly been the cause.

Charon's mass turned out to be a sizable 12% of Pluto's, making Pluto and Charon more of a double-planet system than a massive central object with a slight by comparison orbiting body. In the case of Pluto and Charon, both objects orbit the center of mass of the system.

In 2005, the Hubble Space Telescope sent back photographs of Pluto and Charon showing two more moons (Figure 2). These moons are extremely faint and very distant from Pluto (roughly 5,000 times fainter than Pluto and three times as far from Pluto as Charon).

The moons were named Nix and Hydra. Nix originally gleamed somewhat brighter than Hydra and was thought to be bigger by about 20%. They're now thought to be similar in size. The two moons orbit roughly 27,000 miles (43,000 kilometers) from Pluto.

Figure 2

Images captured by the Advanced Camera for Surveys showed the three moons to have a more neutral color than Pluto's slightly reddish tinge. Pluto's reddish hue probably comes from the interaction of sunlight with nitrogen and methane ice on Pluto's surface. Spectra reveal that Charon's surface is covered mainly with water ice. Given their similar color, Nix and Hydra probably have the same frozen H_2O surface. The theory is that because the three moons share similar color and all orbit Pluto in the same plane, they were all formed by colliding with Pluto in the same crash about four billion years ago. Material blasted off Pluto would have produced the three moons.

Hidden from Sight, Revealing New Information

During the five-year period from 1985 to 1990, Charon's orbit around Pluto was oriented in a line with Earth so that the two objects obscured each other fully (when Charon was behind Pluto) or partially from our view. The phenomenon may have hidden the objects, but it also served as a good opportunity to learn more about them. Measurements showed Pluto to be 2,300 kilometers in diameter and Charon 1,000 kilometers. The two objects are separated by about eight Pluto diameters. For comparison, the Earth is separated from our own Moon by 30 Earth diameters. These data add weight to the argument that Pluto and Charon should be classified as a double-planet system.

Odd Orbit, Changing Atmosphere

Pluto was always an anomaly among planets. Its orbit is highly inclined and very eccentric. It's also retrograde. "Chaotic" is a term often used to describe Pluto's journey around the Sun. In 1989, 10 years before and after its perihelion (the closest point to the Sun in its orbit), Pluto even strayed into Neptune's orbit. Dynamic changes in the objects' gravitational forces prevent them from colliding.

In 1988, Pluto was seen occulting a star. If it had no atmosphere, the star would have quickly winked out. But as we discussed with Triton, when the occulting object has an atmosphere, the starlight dims slowly and minutes go by before it gleams brighter again. The event allows scientists to determine the object's temperature, pressure, and density.

Pluto's surface pressure is a truly miniscule fraction of Earth's: 1/100,000th. At a certain altitude, however, there was a notable change in the fluctuation rate of the star's light. Some phenomenon was taking place at that height. Below that point there was either an atmospheric inversion or a layer of haze. Most of Pluto's atmosphere is composed of molecular nitrogen, probably mingled with some methane. Pluto's surface contains methane ice.

It took 14 years for Pluto to once again occult a bright enough star for useful observations. Even then, knowledge of Pluto's position was too imprecise to pinpoint where the occultation could be observed to within a few hundred kilometers. What was known was that because the disappearing star was so distant, all its light rays were basically parallel, meaning that even from a distance of three billion miles, the shadow cast by Pluto on the Earth would be the same size as the remote object itself. The width of that field was 2,400 kilometers: narrow enough to be easily missed.

Fortunately, the predictions were very well matched to the astronomical event, so it proved a great opportunity to learn more about Pluto. As it turned out, Pluto's atmosphere had expanded to some degree, its pressure had increased, and its temperature, as far from the Sun as it is, was a few degrees higher. Pluto, like Triton, had experienced global warming.

In 2002, Pluto was roughly 13 years past its perihelion, and thus was receding from sunlight. Theoretically, it should have been cooler rather than warmer. The warmer temperature may be analogous to the effect (thermal lag) we experience on Earth when the temperature peaks a few hours after noon. The temperature rise may have resulted from the changing angle of the Sun shining on Pluto's icy landscape. Pluto's atmosphere was still retaining its heat as it continued its 250-year journey around the Sun.

More recent observations of Pluto from occultations show that the temperature still hasn't dropped, though the increase has reached a plateau. That's actually a positive sign for the New Horizons mission, scheduled to approach the Pluto system in 2015. For a journey that was virtually unfathomable until the 20th century was nearly over, it was feared that the atmosphere would cool down to the point that its consituent molecules would condense and freeze, forming essentially snow and the opportunity to examine it would be lost. On Earth, only water freezes and snows out of the atmosphere. On Pluto, it is possible that the entire atomsphere could disappear in a "blizzard" on this remote world. New Horizons has been traveling since January 19, 2006.

Pluto Demoted

Pluto's many peculiarities made it a hot topic since its discovery. Even when it was first recognized as a planet, it was questionable whether it was massive enough to be the object responsible for the perturbations in Uranus's orbit.

Then there was Pluto's chaotic orbit. Its elliptical orbit is so exaggerated that its longest diameter is 25% greater than its shortest diameter. What's more, its orbit is inclined 17 degrees: far surpassing even Mercury's 7 degrees (no other planet's orbit is inclined more than 4 degrees).

Finally, there was the discovery of Charon, and the discovery that Pluto's mass was a puny 1/400th of Earth's. At that point, it became difficult to argue that Pluto was a planet of the same stature as the other eight.

As early as 2000, Pluto disappeared from the planetary display when the new Rose Center for Earth and Space replaced the Hayden Planetarium at the Museum of Natural History in New York City. Director Neil deGrasse Tyson had demoted Pluto on the authority of the institution.

Pluto's proponents claimed the demotion defied history; Pluto had been a planet for 70 years. Then there came another undeniable challenge to Pluto's position: an object larger than Pluto orbiting in the outer realm of the solar system.

This new object appeared in a search of the outer solar system by Mike Brown, a professor of planetary science at Caltech. He extended his search beyond the band closest to the ecliptic, where the other planets orbit, and scanned regions of space high above and below the ecliptic plane. And there it was: the object that outclassed Pluto in size.

The object was 97 AU away, more than twice Pluto's distance from the Sun, and high above the ecliptic. We'll learn more about that object in the next chapter. The critical point was that however slight the difference in size (and with estimates there was still a chance it could be the same size as Pluto or even smaller), it seemed to be bigger than Pluto.

Solar System Scoop

The first official name of the object larger than Pluto was 2003 UB313. This is a provisional designation issued by the International Astronomical Union's Minor Planet Center. 2003 refers to the year of discovery, U denotes the half-month of its discovery (October 16–31), and B313 is its order of discovery in a very busy month for new discoveries.

Now, here's where we get B313 in detail. The letter is chosen from a sequence that progresses through the capital letter alphabet and indicates its order of discovery within the half-month in question. Twenty-five (25) letters are used in this sequence, ("I" is excluded because it could be confused with "1"). the letter is followed (possibly) by some numbers, which indicate the order of discovery within that half-month if more than 25 objects have been discovered. So to figure out the order, you take 25 times the number and add it to the letter's numerical place in the alphabet. In the case of B313 we get: 313*25 + 2 or the 7,827th object designated as a discovery during October 16–31, 2003. This dizzying number shows what our current advanced observing technologies are capable of achieving. It is hard to imagine that so many small objects remain undiscovered in our Solar System, but we find a lot with modern techniques.

After the new object was discovered and its orbit verified, the International Astronomical Union was entrusted with the task of assigning it an official name. Usually, they chose the name suggested by the person who discovered it, in this case, Mike Brown. The quandary was which IAU committee should assign the name; there were separate committees for planets and smaller bodies. The question prompted the designation of another committee—to define what a planet is.

Though it seems improbable, there had never been a formal definition of "planet." For centuries it was intuitive; everyone "knew" what a planet was, or so they thought. We now know that the outer solar system beyond Neptune and Pluto is populated by thousands of orbiting objects. At least six are known to be bigger than Charon, and there could be hundreds bigger than Pluto. We could wind up with a daunting number of planets.

The committee was so divided on the issue that there was no definitive answer. They recommended creating a category of "dwarf planets" that would include objects big enough to be round. That recommendation was taken up at the triennial General Assembly of the IAU held in Prague in 2006. The resolution was proposed at the start of the two-week conference, discussed at meetings, and held up for vote at the closing session.

It was far from the usual meeting where members voted on prosaic technical matters like timekeeping systems. No one expected the degree of controversy and divisiveness the question of what was a planet (or not) provoked. A statement was added to the proposed definition that beyond being spherical, the object "cleared its neighborhood," introducing yet another undefined term. Broadly, it means that the object had enough mass for its gravity to keep similar objects out of its orbit. That would leave Pluto out. At the same time, Pluto crossed massive Neptune's orbit, and even giant Jupiter had Trojan asteroids 60 degrees in front of it and behind it.

Finally, after a lot of discussion (or argument), a vote was taken and a new definition was passed. Planets and dwarf planets were both spherical, but only planets cleared their neighborhoods. A motion was raised to preface the typical planets with "classical," creating two classes of planets, classical and dwarf. That motion failed, so we have planets and dwarf planets. (Ironically, dwarf stars are types of stars and dwarf galaxies are small galaxies, but by the new definition, dwarf planets are not planets.)

Objects meeting the classification of dwarf planet included Pluto, Charon, and the new object, and also Ceres, the largest asteroid. To some conference attendees, Pluto actually earned a promotion. Instead of being a small, remote planet it was the premier object in its new class. Most interpreted it as a demotion, and many were highly critical of the decision.

The change in Pluto's status divided the general public as well as the astronomical community. There have been numerous online positions to overturn the decision. The state of New Mexico passed a resolution honoring their longtime resident, Clyde Tombaugh, and proclaiming that Pluto would always be a planet while in the skies of that state. They declared March 13, 2007 to be Pluto Planet Day. In Illinois, where Tombaugh was born, the State Senate passed a similar resolution in 2009, stating that Pluto was "unfairly downgraded" in status.

In 2007, a group of dissenters met at the Johns Hopkins University Applied Physics Laboratory in Maryland. The meeting included several important planetary scientists, along with Clyde Tombaugh's daughter. Many were involved in attempts to reverse the resolution.

In an unexpected twist of events, the IAU subsequently established a new category of dwarf planet called *plutoids*. Plutoids refer to the dwarf planets beyond Neptune (all the dwarf planets except Ceres, which is in the asteroid belt). So Pluto does have some special status in our solar system. But the IAU resolution stands, and it is no longer a planet.

New Horizons En Route to Pluto

In 2006, NASA launched New Horizons, sending it en route to the Pluto system. The spacecraft was built and operated by the Johns Hopkins University Applied Physics Laboratory. Its principal investigator Alan Stern (then with the Southwest Research Institute) is one of the scientists who decried Pluto's demotion.

New Horizons got a gravity boost from Jupiter in 2007. It flew by Saturn in 2008 and Uranus in 2011. The spacecraft has been sending back images of the Jupiter system, along with a few tests (Figure 5), but it will not be close enough to provide detailed images of Pluto until the summer of 2014. As of March 13, 2013, it was roughly 6.80 AU from Pluto (about 26.07 AU from Earth).

Following its closest approach to the Pluto system (or Pluto-Charon system) in July 2015, the spacecraft will travel farther out in the Kuiper belt. It is expected to be directed toward a Kuiper belt object (one of the many objects orbiting beyond Neptune). The target object will be selected at a future date.

New Horizons launched at the greatest speed ever for a spacecraft. Fairly small in size, it's replete with an impressive suite of equipment. There's Ralph, a visible and infrared imager and spectrometer; Alice, an ultraviolet imaging spectrometer; REX, the Radio Science Experiment; LORRI, the telescopic imager; SWAP, the Solar Wind Around Pluto spectrometer; PEPSSI, the Pluto Energetic Particle Spectrometer Science Investigation, which measures ions escaping from Pluto's atmosphere; and SDC, the Student Dust Counter, constructed by students to measure dust. These various instruments with the colorful names will be bringing us exciting discoveries from the far reaches of our solar system.

Satellite data

Satellite	Mean radius (km)		Mean density (g/cm³)		Geometric Albedo	
Charon	603.6 ± 1.4		1.678 ± 0.151		0.372 ± 0.012	
Nix	44.0 ± 5.0		0.5 ± 0.5		0.08	
Hydra	36.0 ± 5.0		5.0 ± 5.0		0.18	

Sat.	Orbital radius (km)	Eccentricity	Inclination (°)	Period (days)
Charon	17536.	0.0022	0.001	6.387
Nix	48708.	0.0030	0.195	24.86
Hydra	64749.	0.0051	0.212	38.20

CHAPTER 20

Dwarf Planets: Eris, Haumea, Makemake, and Ceres

In This Chapter

➤ Ceres, only asteroid dwarf planet

➤ The Oort Comet Cloud and the Kuiper Belt

➤ The search for dwarf planets

➤ From 2003 UB313 to Eris

➤ Haumea, Hawaiian god

Mike Brown, the astronomer who discovered the object larger than Pluto, was one of the scientists who welcomed the IAU resolution. Brown's take on the controversial decision was thus: "Through this whole crazy circus-like procedure, somehow the right answer was stumbled on. It's been a long time coming. Science is self-correcting eventually, even when strong emotions are involved."

Strong emotions were indeed involved. But once the new object was found, there was no real logic involved in insisting the smaller Pluto was a planet but the larger body was not. The "dwarf planet" category filled the need for a classification of objects that fit Pluto and similar denizens of the outer solar system.

To date, there are five dwarf planets recognized by the IAU: Eris, Haumea, Makemake, Ceres and, of course, Pluto. By definition, they have sufficient mass for their own gravity to have made them spherical and they "clear their orbits." Less massive objects lack the gravitational pull to mold themselves into a round shape. That part of the definition is clear. Clearing their orbits is still imprecise, but the general idea is that they have enough mass to keep large objects out of their orbital paths.

Ceres: Asteroid and Dwarf Planet

In 1801, Italian astronomer Giovanni Piazzi discovered an object he thought was a planet. He observed it repeatedly, but the Sun's glare prevented him from tracking the object long enough to determine its orbit. Fortunately, mathematician Carl Friedrich Gauss had devised a formula for calculating an object's orbit from a few observations. The object was located between Mars and Jupiter.

Piazzi's discovery was quickly followed by a succession of similar orbiting objects. The original object was named Ceres (1 Ceres, as additional objects were found) after the Roman goddess of the harvest. Ceres falls just short in brightness of being visible to the naked eye. Ceres and its companion objects fit Bode's law, which was credible at the time. A more credible source, Johannes Kepler, had observed the gap between Mars and Jupiter in 1596.

Ceres has a diameter roughly 590 miles (950 kilometers). In images captured by the Hubble Space Telescope, Ceres is unmistakably round (Figure 1). It was the first asteroid discovered and is still the only asteroid known to be round. That makes it the only asteroid that fits the definition of a dwarf planet. Extending its unique status, Ceres is also the only dwarf planet in the inner solar system. All the other asteroids are far out in the Kuiper belt.

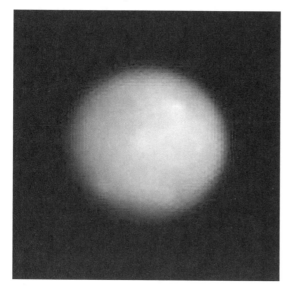

NASA's Dawn spacecraft was launched in 2007 and is scheduled to reach Ceres in 2015. Dawn orbited asteroid 4 Vesta from July 15, 2011, to September 5, 2012, before continuing its journey to Ceres. It should arrive at its destination five months before New Horizons reaches Pluto, beating it out as the first spacecraft to study a dwarf planet in close proximity.

The Kuiper Belt

In the mid-twentieth century, planetary astronomy had lost its allure. Gerard Kuiper, who had emigrated to the U.S. from the Netherlands, was one of the few scientists to buck that trend.

Kuiper's Dutch compatriot Jan Oort had previously discovered a vast cloud of budding comets surrounding the solar system. Named for its discoverer, the Oort Cloud begins way, way out in the solar system, beyond 100,000 A.U. At that great distance, the comets that emanate from the cloud are long-period comets (more than 200 years) that head toward the Earth from all directions.

Kuiper was aware that there were also numerous comets traveling toward Earth from a much closer distance, just beyond Neptune. Their lesser angle of inclination with respect to the plane of the planet's orbits signified that these comets (essentially short-period comets) originate from a belt rather than from a cloud.

Solar System Scoop

Kuiper was not the only astronomer to realize the existence of the asteroid belt. In England, Kenneth Edgeworth arrived at a similar conclusion to Kuiper. The region is also known as the Kuiper-Edgeworth belt and in Europe in particular, the Edgeworth-Kuiper belt.

Eons ago when our solar system was young, the Kuiper belt region was densely packed with enough gas and dust to allow small bodies called planetesimals to form there. In contrast, gas and dust in the Oort cloud region was sparse. The theory is that the powerful gravity of the giant planets forced some of the Kuiper belt's planetesimals out into space, where they formed the Oort Cloud, and even more of them completely escaped the Sun's gravity.

Beginning in the 1990s, advances in observational techniques drove discovery of a plethora of Kuiper Belt objects. Jane Luu and David Jewitt of the Institute for Astronomy of the University of Hawaii capitalized on the combination of superior technology and the exceptional observing conditions at Mauna Kea to search for Kuiper belt bodies.

In 2002, an object was found that turned out to be larger than Charon. The Hubble Space Telescope confirmed its substantial size. The object was named Quaoar (kwa-whar) after the indigenous people who once lived in the area where Los Angeles is today. Quaoar proved more reflective than initially thought, with an albedo of 12%. Another Kuiper belt object, reddish-tinged Sedna, is even bigger than Quaoar.

Searching for Dwarf Planets

After first observing the object 2003 UB313 (the one that led to Pluto's demotion), Mike Brown led the quest to find other large objects orbiting in the outer solar system. Since the new object was twice as distant as Pluto, at 97 AU, and high above the ecliptic, as we learned in the last chapter, there were sure to be other such objects in the unexplored nether reaches way beyond Neptune.

Solar System Scoop

Reasonably high-resolution images can be used to measure an object's size. At the distance of the Kuiper belt objects, even Hubble images are fuzzy. To compensate, the sizes of such remote objects are often derived from a presumed albedo. Calculations reveal the size that would produce the brightness observed from infrared images.

The official names of celestial objects are assigned by the International Astronomical Union. Mike Brown gave his discoveries nicknames. The largest object was Xena, after the warrior princess of the popular television show. Its satellite became Gabrielle, Xena's sidekick.

When names were assigned the dwarf planets, 2003 UB313 became Eris (Figure 2). Like Xena, Eris has its own satellite, named Dysnomia (Figure 3). Dysnomia was the Greek goddess of lawlessness. Though there is no official connection, Lucy Lawless was the actor who played Xena in the eponymous television show.

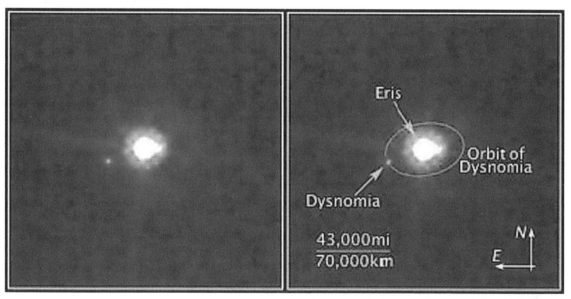

Figure 2

Figure 3

Makemake

Makemake (mak-ay-mak-ay) is the third largest Kuiper belt object, after Eris and Pluto. Discovered in 2005, Makemake was classified as a plutoid in 2008. It was one of the discoveries made by Mike Brown and his team of David Rabinowitz of Yale and Chad Trujillo of the Gemini Observatory in Hawaii. Discovered at Easter time, it got the name Easterbunny. Makemake, as it became formally known, is an Easter Island god.

Makemake is about one-sixth as bright as Pluto is. It can be seen with an amateur telescope, but only as a tiny point of light (Figure 4). Unlike Pluto, Eris, and other large Kuiper belt objects, Makemake has no moon that would help us learn more about it.

Makemake is approaching its aphelion, its orbit's farthest position from the Sun (52 A.U.). Its journey around the Sun takes 310 Earth years. Its fairly steep orbital inclination—29 degrees—is one reason it was not discovered earlier.

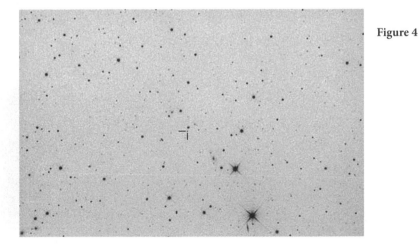

Figure 4

Haumea

In late 2004, Mike Brown claimed that he would discover a planet bigger than Pluto by January 1, 2005. He even made a bet over the bold claim. A few days before the New Year, the team of Brown, Rabinowitz, and Trujillo spotted a likely candidate. Brown lost the bet—the object was one-third Pluto's size. However, it proved an exciting discovery nonetheless. Perhaps predictably, the new object was nicknamed Santa.

A graph of the object's brightness revealed tremendous fluctuations in its light curve. Shaped like a cigar, it tumbles around every four hours. The curious object has the fastest spin of any large object in the solar system. It spins so fast that even though it has enough gravity to be round, it gets pulled out of shape by its own speedy rotation. Because its gravity still accounts for a lot of its shape, it classifies as a dwarf planet by the IAU's lax definition.

Brown named the object Haumea, for the Hawaiian goddess of fertility and the patron of the Big Island, home of the Mauna Kea observatory. It has two tiny moons, Hi'iaka for the patron goddess of the Big Island, born from the mouth of Haumea, and Namaka, a water spirit born from Haumea's body.

Haumea is an ellipsoid measuring 1,960 kilometers x 1,518 kilometers x 996 kilometers based on the assumption of an albedo of 73%. However, those figures are estimates that may not be accurate. Theoretically, Haumea's size could be gauged during a stellar occultation.

Beginning in 2008, there were predictions of a series of mutual occultations of Haumea and its satellite Namaka. Teams of astronomers located throughout the globe strived to detect the occultations (or transits or shadow transits), but the drop in brightness was so minute that it eluded them. The Hubble Space Telescope might be more successful in that endeavor.

Quite a few Kuiper belt objects of various sizes share Haumea's spectra. A reddish region was observed in 2009. It might have been a culprit in complicating the efforts to detect the mutual occultations.

Haumea is roughly 15 times fainter than Pluto. It's bright enough to be seen by experienced amateur astronomers (Figure 5).

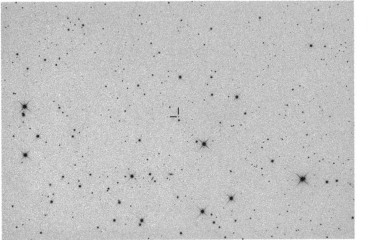

Figure 5

More Dwarf Planets

Since the category was formed, the number of dwarf planets has been increasing, and promising candidates continue to be observed. They're also called TNOs, for Trans-Neptunian Objects.

Some of the scientists attending the 2006 IAU conference wanted the definition of dwarf planet to contain a size criterion. They lost that vote, but based on the objects we know of, we can infer the size needed to have sufficient mass to be round. There may be a lower threshold for icy objects and a higher threshold for rockier ones.

Moons are valuable for determining the mass of the central object, and a good proportion of the Kuiper belt objects have moons. For example, Eris is roughly 25% more massive than Pluto, though only 5% bigger.

Ultimately, the Kuiper belt may turn out to be populated by hundreds of objects that classify as dwarf planets (or plutoids or TNOs—take your choice).

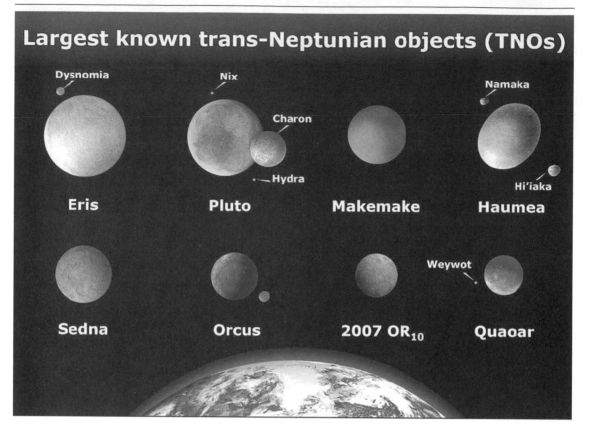

Figure 6

Solar System Scoops

Dwarf Planets by the Numbers

KBO	aphelion	perihelion	diameter	period	eccentricity	inclination
Pluto	49.3	29.7	~4,600 km	248 yrs	0.25	17°
Eris	97.6	37.8	~2,600 km	557 yrs	0.44	44°
Makemake	53.1	38.5	~3,200 km	310 yrs	0.16	29°
Haumea	51.5	35.1	~3,000 km	285 yrs	0.19	28°

PART FIVE

Eclectic Objects

CHAPTER 21

Comets: Bright "Hairy Stars" and Their Tails

In This Chapter

➤ Halley's dedication and diligence
➤ Icy conglomerates (or dirty snowballs)
➤ Periodic and non-periodic comets
➤ Discovering comets from space
➤ Into the nucleus of a comet

When the ancients looked up at the sky, they were used to seeing the Sun predictably rise and set, the Moon wax and wane, the stars dazzle brightly, and the "wanderers" journey across the sky in their usual path. Then abruptly, perhaps every 10 years or so, a bright object would streak by, seemingly out of nowhere, and then vanish from sight. The object looked like a star with hair…and a tail.

The word "comet" comes from the Greek word for "long-haired." Aristotle described the objects he saw as "stars with hair." Centuries later, Fred Whipple of Harvard identified the "hairy stars" as "dirty snowballs." The "hairy stars" (or "falling stars" in some ancient legends) are bodies made up of various ices with dirt and chemicals thrown into the mix.

Comets appear to be streaking because of their tails, but in reality they rise and set with the stars. They may move slightly from one night against the milieu of stars, so they appear different from Earth, but they rarely noticeably change shape or position within a short time period, like on an hourly basis.

Though the word "tail" immediately comes to mind when one imagines a comet, not all comets have huge, trailing tails. Nor are all comets especially bright. Many are actually dim and unremarkable. Especially in our light-polluted environment, they pass by virtually unnoticed by most of the general public.

On the other hand, some comets are outstanding. Comet Hale-Bopp (for its almost simultaneous discovery by Alan Hale and Thomas Bopp) shone brilliantly in the sky in the 1990s. It reached its brightest moment in 1997, where it was visible even throughout heavily lit New York City. For several weeks, it could be easily seen just by looking up at the sky.

Solar System Scoop

We went over the alphanumerical system scientists use to designate new celestial discoveries. Scientifically, Comet Hale-Bopp was given the designation C/1995 O1: C/ for comet, 1995 for the year it was discovered, O for the half-month of its discovery, and 1 to denote that it was the first comet discovered that half-month.

Comet Hale-Bopp not only had one gleaming tail, but two (Figure 1). The brighter "dust tail" was whitish and curved gently back from the comet's head. The fainter "gas tail" was bluish and straighter. The dust tail is composed of the dust cast off as the comet makes its way around the Sun. The brightness we see is sunlight reflected by dust. The bluish tail is gas emitted by the comet and thrown outward away from the Sun's direction by the solar wind. In fact, the solar wind was initially found to be the reason that comet tails always point away from the Sun, even when the comet is traveling in the same direction.

Figure 1

As of January 2011, there were 4,185 known comets. The number keeps increasing, but it's only a miniscule fraction of the potential number of comets in our solar system. The Oort Cloud alone could be home to one hundred billion comets.

From Bad Omen to Edmond Halley and the Age of Enlightenment

Ancient texts from various cultures contain references to the sudden appearance of comets. They were usually seen as bad omens, predicting death, disaster, or defeat. Halley's Comet appeared in 1066, just before the Norman invasion of England. If it was a bad omen for the English, it was a good omen for the conquering Normans. The comet's appearance—and the Norman conquest of England—is immortalized in the Bayeux Tapestry, which shows King Harold II being informed of the comet before the Battle of Hastings.

Aristotle recognized that comets were not planets or related to planets, as some earlier philosophers claimed. Unlike the planets, which followed a predictable course, comets showed up haphazardly in any part of the sky. But Aristotle's theory of comets was wrong in another respect. He argued that comets were a phenomenon of the Earth's upper atmosphere.

Others disputed that viewpoint. After all, if comets could travel through any part of the heavens why would they be restricted to Earth's atmosphere? As with the geocentric theory, however, it was Aristotle's view that prevailed. Finally, in 1577, Tycho Brahe used measurements taken from a bright comet that was seen over several months to calculate that the comet was more than four times farther from Earth than our own Moon. It was definitively way beyond the Earth's atmosphere.

Comets appeared so suddenly and sporadically that for centuries no one thought there was any connection between the bright objects. For example, Edmond Halley and Isaac Newton each observed comets in 1680 and 1682. In 1682, gleaming comets were visible in the morning sky and the evening sky, each independently.

Halley was especially interested in comets. As we discussed in Chapter 2, Halley was Newton's champion and drawings of the comet and its tail appeared in Newton's groundbreaking *Principia Mathematica*.

EDMUNDUS HALLEIUS R.S.S.
Astronomus Regius et Geometriæ Professor Savilianus.

Edmond Halley

By the 18th century, Halley had gathered a wealth of data on comet sightings and measurements. He fervently sought out all the material he could find, and attempted to calculate the orbits of comets using Newton's theory of gravity. Applying the theory to 23 comet sightings from 1337 to 1698, he realized that the comets appearing in 1531, 1607, and 1682 had strikingly similar orbits.

Indeed, Halley concluded that it was the same comet returning. Slight variations in the intervals between its last two visits could be explained by Jupiter's gravity.

Most significant, Halley figured out that comets travel in elliptical orbits like planets. Comets follow Kepler's laws of planetary motion. The distinction is that the orbits of comets can deviate from a circular path to a much greater degree than a planet's. Comets often have very exaggerated elongated orbits.

In accordance with Kepler's second law, comets speed up when they move close to the Sun and slow down when they are far away from it. In applying Kepler's second law, the line connecting the Sun and an orbiting body (in this case, a comet) sweeps out equal areas in equal times. That means that the triangle swept out by a line connecting the Sun and a comet is short and wide when the comet is close to the Sun and long and narrow when the comet is far from the Sun. And in the same time frame, a wide swath means the comet has to go fast and a narrow swath means it has to move slowly.

Halley's Comet has a period of roughly 76 years. During most of that time it is far from the Sun and consequently moves slowly. It was not until the advent of the Hubble Space Telescope and giant terrestrial telescopes that the comet, intrinsically faint and very distant, could be followed throughout the full course of its orbit. When it moves closer, roughly within Jupiter's orbit, the solar energy vaporizes some of its ice and the tail begins to emerge. It appears to us through reflected sunlight.

As Halley's Comet makes its way through the inner solar system, its tail grows and it gets notably brighter. It's visible to us for a few months. First it appears in the evening sky, then it disappears as it travels behind the Sun, and then it reemerges again in the morning sky.

Apparitions of Halley's Comet were recorded as early as 240 B.C. It's the brightest comet to appear regularly in our skies. Note the word "regularly." There are some extremely bright comets that show up sporadically. (For a comet to appear dazzlingly bright, it has to be both intrinsically bright and on a trajectory that brings it relatively close to the Earth.)

From calculations he made based on the comet's earlier apparitions, Halley predicted that the comet would return once again in 1758. Halley died before he could find out whether or not he was right. His predictions had been slightly refined, but the presumed year was still 1758. Throughout the year, people all over the world futilely searched for the comet. Finally, at Christmas time, a farmer spotted the comet! Halley's hard work was validated, and Halley is immortalized by the comet that bears his name: Halley's Comet, the most famous comet ever.

Solar System Scoop

In 1847, a woman named Maria Mitchell on Nantucket Island off the coast of Massachusetts, looked upward and discovered a comet. At that time, anyone who discovered a comet received a gold medal from the King of Denmark. It took something of a fight, but Mitchell got her reward. She eventually became a professor of astronomy at Vassar College and was one of the most famous astronomers of the late 19th century.

Dirty Snowballs

In 1950, Fred Whipple began a long and illustrious career as a professor of astronomy at Harvard University. Shortly after, he wrote a series of papers entitled "A Comet Model," in which he presented his theory of comets as "icy conglomerates." Originally published in

The Astrophysical Journal, Whipple's theory caught on with the popular press, where "icy conglomerates" became "dirty snowballs." Ever since then, comets have been known as icy snowballs: ices with dust and chemicals mingled in.

Comets typically orbit in a far distant path from the Sun. Pulled in toward the Sun (possibly by the gravity of a passing star), the intense solar energy eventually vaporizes some of the ice. This direct transformation of solids to gases is called "sublimation." That is, the ices "sublime" rather than melt, which would cause them to liquefy.

As Comet Hale-Bopp illustrates so well, the gas tail is the vapor blown away from the Sun, and the dust tail is the dust left behind in the comet's orbital path.

It was also in 1950 that Jan Oort discovered the vast cloud where these incipient comets were born. The Oort Comet Cloud is an immense sphere spanning out almost a full light year. Its radius is roughly 50,000 times the radius of Earth's orbit, or 50,000 AU. That's more than 1,000 times farther out than Neptune's orbit.

For such a huge expanse, the Oort Cloud contains minimal mass. The total mass of objects populating the Oort Cloud is only three times the mass of the Earth.

Most long-period comets originate in the Oort Cloud. Those who argued against Aristotle's idea that comets are limited to Earth's atmosphere pointed out that the bright objects they saw were coming from anywhere in the sky. The fact that comets come in from the Oort Cloud at all angles led to the conclusion that it is spherical in shape.

Eons ago, the potential comets populating the Oort Cloud might have been where the giant planets now orbit the Sun. As the orbits of the giant planets moved outward, the small objects were thrown even farther out to form the huge cloud.

Short-period comets orbit closer to the plane of our galaxy. As we discussed in the last chapter, these comets originate in the Kuiper belt. Kuiper belt objects can be cast into very elongated orbits through gravitational interactions with the giant planets, which send them into the Oort Cloud. Most, however, have much smaller orbits.

The Kuiper belt, spanning from roughly 30 AU to 50 AU, is thought to be home to hundreds of thousands of orbiting bodies. Pluto and Eris are currently both regarded as large Kuiper belt objects. Innumerable smaller objects are budding comets.

Colliding with Jupiter

Independently and working collaboratively with her brother, Caroline Herschel discovered eight comets in the late 18th and early 19th century. Much more recently, astronomer Caroline Shoemaker, her husband geologist, Eugene Shoemaker, and amateur astronomer David Levy discovered 30 comets. As skilled as William Herschel was at building telescopes, they were no match for the Palomar Observatory's wide-angle field telescope used by the Shoemakers and Levy.

The most famous of their collaborative comet finds was number nine. Comet Shoemaker-Levy 9 was discovered on March 24, 1993, orbiting Jupiter. It was the first comet observed orbiting a planet. That distinction, however, was not what gave it its fame. The comet had broken up into pieces and the fragments were headed right toward Jupiter! Probably captured by Jupiter 20 or 30 years earlier, it broke up during a close encounter with Jupiter in 1992, and was headed back after completing its orbit.

No one had ever directly observed extraterrestrial objects colliding. From the ground and from space, telescopes of all types tracked the fragmented object making its way toward its target. The Hubble Space Telescope in particular, kept watch. In mid-July 1994, the awaited event finally happened: Comet Shoemaker-Levy 9 hit Jupiter.

Terrestrial telescopes of all sizes were aimed at the sky attempting to capture the awesome event. Powerful telescopes could pick up the myriad bits of the comet hitting the giant planet, one after another, after another (Figure 3). Dark material—hydrocarbons—spewed out from under the Jovian cloudtops. The airborne particles were an eclectic mixture: material from the incoming comet, a fusion of particles from the comet and particles from Jupiter's atmosphere created by shock waves from the powerful impact, and material from the lower levels of Jupiter's atmosphere churned up by the fireball that followed the crash.

Jupiter was left with huge scars that were visible for months after the massive collision. By some accounts they stood out even more than the Great Red Spot. Eventually, they were broken up by shears from Jupiter's cloud band.

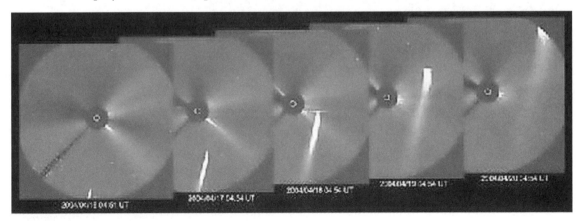

Figure 3

Comet Shoemaker-Levy 9 was not the only dirty snowball to hit Jupiter. On July 19, 2009, an amateur astronomer spotted a new impact site on the giant planet. The next day, NASA's Infrared Telescope Facility on Mauna Kea captured it as a bright region marking Jupiter's surface (Figure 4). The culprit was most likely a comet nucleus about a mile or so in diameter.

The 2009 impact was totally unexpected. These events spur apprehension about the prospect of giant impacts occurring on Earth. Projects like Pan-STARRS (which we'll get to soon) are busy scanning the skies to detect any objects that that could potentially wind up in a major collision with our terrestrial home.

Figure 4

Periodic and Non-Periodic Comets

About 200 comets are known to have periodic orbits, ranging from 3.3 years for Comet Encke to thousands of years. According to Kepler's law, the average size of the orbit is related to the period of the object. In other words, Encke's comet, returning in just a few years, has a much shorter orbit than Halley's Comet, with a period of 75.3 years.

Halley's Comet has an extremely eccentric orbit (97%), varying from 60% of the distance from the Earth to the Sun to 35 times that distance (0.59 AU to 35 AU). In addition, its inclination is 162 degrees, meaning it orbits backwards from almost all other solar system bodies. (The comet's odd backward orbit was one of the clues Halley picked up on in connecting the three apparitions.)

Encke's comet was originally discovered in 1786, but it was not recognized as a comet until 1819, when Johann Franz Encke calculated its orbit. Encke's comet varies in its orbital path from 0.33 AU to no more than 4.11 AU. Its average length is only 2.2 AU. We will be able to see it again in the fall of 2013. For Halley's Comet, we have a much longer wait. It won't be returning until about a year before its perihelion, which happens on July 28, 2061.

Most comets are named after the people who discovered them. Halley's Comet defies this rule by default since there is no record of who first discovered or sighted it millennia ago. Encke's comet, though, is unusual because it was first seen by Pierre Méchain, who recorded his observation, but instead named after the person who calculated its orbit. Currently, the IAU is responsible for evaluating the initial report and assigning the comet a name. Up to three people can share the name of a comet they jointly discovered.

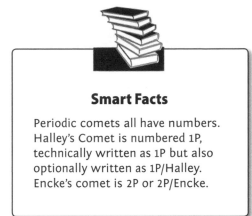

Smart Facts

Periodic comets all have numbers. Halley's Comet is numbered 1P, technically written as 1P but also optionally written as 1P/Halley. Encke's comet is 2P or 2P/Encke.

Historically, most comets were discovered by a lone person. Recently, there have been some huge automated projects sweeping the sky for comets. These include the **L**owell **O**bservatory **N**ear-**E**arth **O**bject **S**earch (LONEOS); **N**ear-**E**arth **A**steroid **T**racking (NEAT) with bases at the Palomar Observatory in California and at Haleakala on Maui, Hawaii; and MIT's **L**incoln [Laboratory] **N**ear **E**arth **A**steroid **R**esearch (LINEAR), based in Socorro, New Mexico. These ambitious projects have discovered numerous comets on the list of periodic comets and even more non-periodic comets.

Currently, only LINEAR is still in operation. As its name and others imply, their searches include asteroids as well as comets. A top priority is searching for asteroids and comets that might venture close to the Earth.

Solar System Scoop

Pan-STARRS, for the Panoramic Survey Telescope and Rapid Response System is a cutting-edge project of the University of Hawaii equipped for discovering vast numbers of comets, asteroid, and variable stars, among other celestial bodies. Its main mission is seeking out close objects that threaten to hit the Earth.

Comet PANSTARRS (C/2011 L4), discovered by the Pan-STARRS array, is a non-periodic comet that first appeared in June 2011 and made a spectacular sight in March and April 2013 (even lingering into May and June). It reached its perihelion on March 9th and dazzled observers in North America.

SOHO in Space

In 1995, the European Space Agency launched the Solar and Heliospheric Observatory (SOHO). The spacecraft was sent out about one million miles, 1% of the distance to the Sun, where the gravity of the Earth and Sun balance. On board is an international array of instruments, from NASA, the U.S. Naval Research Laboratory, and other space and research agencies.

SOHO floats in that region of space, relying on jets of gas to maintain a "halo orbit," a slight circle that keeps it from straying into a path that would disrupt radio telescope signals. Its instrument suite includes a pair of coronagraphs: telescopes that block out Sun's normal surface light and allow the surrounding solar corona (about a million times fainter) to be seen. C2 (the second coronagraph) shows a ring of space out to a distance a few times the radius of the Sun. C3 has a wider field of view. (C1, the innermost coronagraph has stopped working.)

SOHO's coronagraphs routinely see comets traveling toward the Sun. Comets that close to the Sun are invisible to terrestrial telescopes because the Sun's glare obscures them from view. SOHO picks up comets that are not only close, but in some cases actually hit the Sun (Figure 5).

SOHO has found more than 1,000 comets. Most of the "SOHO comets" are part of a comet family known as Kreutz sungrazers. Traveling in similar orbits, they are fragmented remnants of a primordial object that orbited in the outer realm of the solar system. They are definitely long-period comets—500 to 1,000 years—but they are assigned periodic comet numbers.

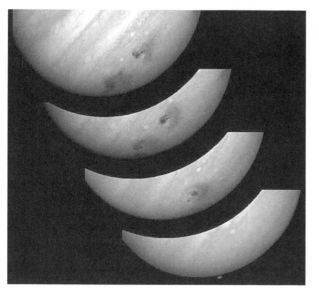

Figure 5

Sudden Brilliance

Among the innumerable comets in the solar system, only Halley's Comet is predictable far in advance. Some comets are known a few months in advance; others for just a few days.

In 2007, Comet 17/P Holmes, a periodic comet discovered over a century earlier, abruptly became vividly bright. Improbably, it grew a million times brighter with a few days. The comet appeared every seven years, but this was the first time it shone conspicuously bright to anyone who chanced to look up at the sky.

Comet Holmes appeared as a fuzzy ball that grew consistently bigger from day to day. Old records revealed that it was not actually the first time that happened; it had just as suddenly brightened in 1892. In 2007, we had much better technology. Cameras aloft in space zeroed in to study the phenomenon in detail.

The sudden brightening was attributed to a huge cloud of dust cast off by the comet. As it grew in size, the dust cloud reflected more sunlight. The comet was virtually all head and no tail. Its stumpy tail broke off and drifted out into space. For a few days, though, the unexpected comet was quite a remarkable sight.

Comets Up Close

Since the 1980s, space expeditions have been traveling to comets, able to enter the comet's head and get close to the nucleus. The first missions included a fleet of spacecraft from the European Space Agency, the Soviet Union, and NASA, ready to capture a sojourn of Halley's Comet to the inner solar system in 1985–86. The premier spacecraft was ESA's Giotto, named for the Italian painter. Giotto had seen the 1301 apparition of Halley's Comet. It appears as the Star of Bethlehem in a fresco the artist created for a chapel in Padua.

Photographs taken by Giotto revealed that the nucleus of the illustrious comet is shaped like a potato. Its longest dimension extends roughly 10 miles (16 kilometers), and its shorter dimension is roughly five miles (eight kilometers). It's extremely dark; no more than 2% to 4% of the light hitting it gets reflected. The world's most famous dirty snowball is coated in a dark crust.

Cracks in the comet's crust spew out jets that are mainly water (80%), with particles of other materials mixed in. Giotto directly picked up some of those particles. The gas and dust particles first evolve into the coma (the area surrounding the nucleus) and then the tails.

Each time a comet approaches the Sun, it loses some of its mass. Comets can survive hundreds of trips near the Sun. However, there are also extinct comets that have lost virtually all their active gases and dust and look more like small asteroids.

The Halley's Comet expedition was just the start of the comet missions. NASA's Deep Space 1 traveled to Comet Borrelly in 2001, after testing ion propulsion at asteroid Braille following its 1999 launch.

NASA also launched Stardust in 1999. In 2004, it flew by the volatile 81 P/Wild-2. Using light, sticky aerogel, Stardust trapped some of the comet's dust and brought it back for analysis. It turned out the comet contained an eclectic variety of dust particles and molecules including some from stars way beyond our solar system.

Deep Impact was launched in January 2005. Its target was Comet Tempel 1. On July 3, 2005, the probe impacted the nucleus, sending up a huge bright dust cloud and gouging a crater the size of a football field. It was the first time a space probe had ever directly blasted material from a comet.

Comet Tempel 1 turned out to have more dust but less ice than the scientists expected. They were also surprised at how fine some of the dust was. Images of the nucleus showed a variety of terrains, including smooth areas that must be fairly young.

The Deep Impact mission generated tremendous interest in the general public as well as the science community. The spacecraft is now on an extended mission called Extrasolar Planet Observation and Deep Impact Extended Investigation (EPOXI).

The next destination was Comet Hartley 2. On November 4, 2010, Deep Impact flew within 430 miles (700 kilometers) of the comet and sent back detailed images of it nucleus, along with shots of several bright jets.

From February 20th to April 8th of 2012, Deep Impact's Medium Resolution Instrument captured images of Comet Garradd, using various filters. With each comet visit, scientists learned more about the dirty snowballs and could compare the compositions and other properties. But it was Deep Impact's next venture that raised expectations that we would soon be seeing something truly spectacular.

Rosetta

The most ambitious comet mission is ESA's Rosetta, named after the Rosetta Stone. The Rosetta Stone was inscribed in 196 B.C. with hieroglyphics depicting the same message in three languages: ancient Egyptian, Demotic (another Egyptian language), and classical Greek. Because it was essentially the same text in three languages, Egyptologists two centuries ago were able to decipher the hieroglyphics. The Rosetta spacecraft is designed to allow scientists to decipher comets.

Launched in 2004, Rosetta should reach comet 67P/Churyumov-Gerasimenko in 2014. After getting gravity boosts from Earth in 2005 and from Mars and Earth again in 2007, it flew by asteroid 2867 Steins in 2008, and following another boost from Earth, 21 Lutetia in 2010.

In November 2014, Rosetta's Philae probe will actually land on the comet. (Philae is an island in the Nile where an obelisk was found that helped decipher the Rosetta Stone.) From November 2014 through December 2015, Rosetta will be orbiting Comet Churyumov-Gerasimenko: the first spacecraft ever to orbit around a comet.

CHAPTER 22

Faint Asteroids, Flashing Meteors

In This Chapter

➤ Denizens of the asteroid belt

➤ Imaging asteroids from space

➤ Capturing an asteroid

➤ Meteor showers light the sky

➤ Finding meteorites on Earth

The skies above Earth are populated with an eclectic assortment of objects of various sizes, shapes, motions, and magnitudes. The earliest known, apart from the Sun and Moon, were the stars and planets. So whenever an unfamiliar object appeared, it was typically thought to be a "new star" or a "new planet" or at least, "star-like" or "planet-like."

As we discussed in the chapter on dwarf planets, when Giovanni Piazzi discovered the first asteroid in 1801, he indeed thought he found a new planet. By the mid-1800s, Ceres had lots of company as many more similar bodies were found. Eventually, it became obvious that they were much too numerous and much too small to be reasonably regarded as planets. About 1860, they became known as asteroids (though late into the 19th century, the terms "asteroid" and "planet" were still sometimes used interchangeably).

The term "aster" means "star." Asteroids have no relation to stars, but with 19th-century telescopes, they appeared as faint points of light, similar to stars. William Herschel proposed the term "asteroid," meaning "star-like" or "star-shaped."

Meteors have been called "shooting stars" since antiquity. The brilliant objects streaking across the sky are actually bits of space debris burning up as they enter Earth's atmosphere.

In meteor showers, they flash by every few seconds or minutes and seem to come from the same point in the sky. The Perseid meteor shower in August is known for its extremely impressive display.

Meteors are particles of asteroids or comets. We will begin this chapter with asteroids.

Discovering Asteroids

Ceres, the first asteroid discovered, has the dual designation of being an asteroid and a dwarf planet. There may be other dwarf planets among the myriad asteroids, but so far Ceres is the only one known.

Ceres was followed by three more asteroids discovered in quick succession: Pallas, Juno, and Vesta. Four decades went by before the next cadre of asteroids was discovered. Then the count picked up again as many asteroids were discovered during the 1840s and 1850s.

In 1891, German astronomer Max Wolf, a pioneer in the field of astrophotography, applied the technique to searching the skies for asteroids. Wolf is credited with discovering 248 asteroids.

More than a century later, hundreds of thousands of asteroids have been discovered by sky-mapping programs on the ground and in space. We mentioned some of the ground-based efforts in discussing the search for comets: LONEOS, LINEAR, and NEAT. Other projects that have contributed to the exponential rise in the asteroid count include Spacewatch, a University of Arizona project that scans the skies with a telescope at the Kitt Peak National Observatory; Project Catalina, which uses telescopes in the northern hemisphere in Arizona and the southern hemisphere in Australia; and Japan's Spaceguard.

Asteroids showed up as short streaks on Max Wolf's long-exposure photographic plates. With space mapping, asteroids are especially prominent in the infrared, where their heat generates peak radiation. NASA's Spitzer Space Telescope, in particular, uses infrared mapping to target specific areas of space and examine individual asteroids and comets. From December 2009 to February 2011, NASA's Wide-field Infrared Survey Explorer (WISE) discovered tens of thousands of asteroids.

Most of the asteroids populate the asteroid belt between Mars and Jupiter. However, there are other clusters of asteroids as well. One group recently making headlines are the near-Earth asteroids. Near-Earth asteroids originate in the asteroid belt, but are thrown into the inner solar system by interactions with Jupiter. NASA has plans to capture one of these renegades and drag it into orbit around the Moon.

The ideal candidate for NASA's asteroid mission is between 20 and 30 feet long (32.2 to 48.3 kilometers). Asteroids vary greatly in size. Some are scarcely one meter across. On the other side of the asteroid scale, six asteroids are bigger than 200 miles (300 kilometers) across, and more than 200 are wider than 60 miles (100 kilometers). The masses of all the asteroids added together equals only about 1/1000 the mass of the Earth.

Solar System Scoop

In 1908, a mysterious object exploded in the sky over Tunguska, Siberia in Russia. A huge expanse of forest was demolished. Trees were strewn over the ground, pointed radially away from the blast. It was probably a small meteoroid or a comet; what it was precisely was never determined.

More than a century later, on February 15, 2013, the town of Chelyabinsk in Russia was hit by tremendous shock waves from a 10,000-ton meteor exploding as it hit the Earth's atmosphere. This time, about 1500 people were injured, including 200 schoolchildren.

Both events seem like something out of science fiction, but obviously they are not. Scientists are aware of the dangers presented by near-Earth objects, and in particular, near-Earth asteroids. Astronomers are now charting all near-Earth asteroids: those whose orbits are close to our planet's or even cross it. If an object is speeding toward us and we only detect it days or weeks in advance, nothing can be done to stop it. But detected early enough, it can potentially be diverted.

Imaging the Small Space Rocks

Asteroids are too small and too distant to be seen very well by even powerful ground-based telescopes. Using radar, however, giant Earth-based telescopes have provided valuable details of the shapes of numerous asteroids. The Arecibo Radio Telescope in Puerto Rico is one of the most successful in this venture.

The Hubble Space Telescope has detailed the shapes of some of the asteroids. Hubble images showing that Ceres is round helped earn it dwarf planet status. On the other hand, Vesta, the fourth asteroid found, seems to have an irregular shape.

Tracking the brightness of asteroids over time gives us more clues about their shape. The light curves (graphs of an object's brightness) of most asteroids show regular fluctuations, signifying the rotation of an object that either has regions of light and dark or else is irregularly shaped.

Phobos and Deimos, the two small moons orbiting Mars, are believed to be captured asteroids. The two moons have been photographed repeatedly by spacecraft dating back to Mariner 9 in 1971. They provide good illustrations of what asteroids look like close up.

Figure 1. A composite photo showing Mars's moon Phobos (lower right), Mars's moon Deimos (lower left), and asteroid Gaspra. Gaspra, observed in 1991 from the Galileo mission en route to the Phobos and Deimos images were taken by one of the Viking orbiters in 1977.

Flying By, Landing on Asteroids

En route to the Jupiter system, Galileo passed asteroid Gaspra in 1991. Gaspra's irregular shape measures roughly 12 miles x 8 miles x 7 miles (19 kilometers x 12 kilometers x 11 kilometers) across. It rotates about every seven hours.

Images of Gaspra show the small rocky object has been subject to a lot of collisions and battering. Its surface is marked by a huge depression roughly four miles (six kilometers) across and a large crater about one mile (1.5 kilometers) across situated on its day-night line. Numerous small craters, including more than 600 ranging from 100 to 500 yards (91 to 457 meters), dot its surface. As another sign of its chaotic history, Gaspra also has grooves that might be fractures. These measure 100 to 300 yards (91 to 274 meters) across.

Two years after imaging Gaspra, Galileo turned its cameras to Ida. Ida turned out to have a tiny moon, Dactyl (Figure 2).

Ida has a rotation of 4.6 hours, which allowed its entire surface to be easily captured by Galileo. Its average radius was calculated at 10 miles (16 kilometers). Minute Dactyl has a radius of only 0.4 miles (0.7 kilometers).

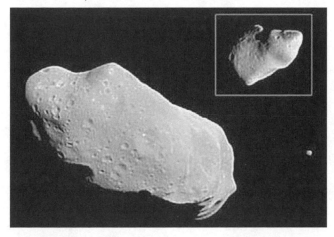

Figure 2. Flying by asteroids was an offshoot of Galileo's main mission. Several years later, NASA sent out an asteroid mission equipped with a probe. It was originally called the Near Earth Asteroid Rendezvous (NEAR). The name was later changed to NEAR Shoemaker in memory of Eugene Shoemaker, a leading pioneer in the field of planetary science (as well as noted geologist and comet discoverer).

In 1997, NEAR Shoemaker flew by asteroid Mathilde and then went on to Eros. First the spacecraft orbited Eros, mapping and studying the asteroid in detail, until it eventually landed on Eros (Figure 3).

Figure 3. Mathilde is an example of a C-class asteroid, meaning it is rich in carbon compounds (C stands for carbonaceous). Consequently, it has an extremely low albedo (think of soot).

Eros, on the other hand, has an albedo six times brighter. Eros, Gaspra, and Ida are all S-class asteroids: S for stony. For the most part, S-class asteroids inhabit the inner part of the asteroid belt and C-class asteroids inhabit the outer part.

NEAR Shoemaker landed on Eros on February 12, 2001. From the asteroid's surface, it was able to capture extremely detailed images until its signal finally died (Figures 4 and 5).

Figure 4

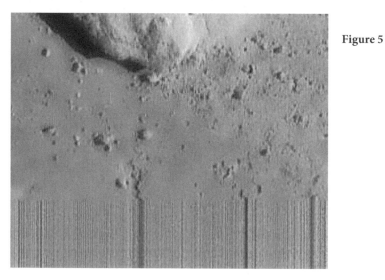

Figure 5

Other missions have been flying by and landing on asteroids. Deep Space 1 flew by Braille in 1999. It subsequently passed Annefrank in 2002. On its way to comet 67P/Churyumov-Gerasimenko, Rosetta passed by asteroids Steins in 2008 and Lutetia in 2010.

JAXA sent the Hayabusa mission to the S-class asteroid Itokawa in 2005. Originally called MUSES-C, its name was changed to Hayabusa, which means "peregrine falcon."

Itokawa is only 540 x 270 x 210 yards (494 x 247 x 192 meters) across. Hayabusa orbited at altitudes of 2 to 12 miles (3 to 20 kilometers) above the asteroid's surface, where it examined its surface, gravity, mineral composition, and other characteristics. Problems occurred with several systems, threatening to compromise a major goal of the mission, which was to bring back samples to Earth for analysis. But despite these problems, Hayabusa returned with particles confirmed to have come from Itokawa.

The particles turned out to be virtually identical in composition to the dust found in meteorites. Based on analysis of the dust, tiny Itokawa probably broke off from a much larger asteroid.

NASA's Dawn has completed its study of Vesta and is now heading for Ceres. Dawn went into orbit around Vesta on July 16, 2011. The results are still being analyzed. but one thing is certain: Vesta is truly ancient. NASA scientists propose that Vesta is the "last of its kind"." That is, it's the only surviving example of the large planetoids came together to form the terrestrial planets at the dawn of the solar system.

Vesta's surface bears irregular dark spots and streaks, most probably the scars of asteroid collisions eons ago. Images also revealed gullies that look like they were eroded by flowing liquid, which could have even been water.

Dawn escaped Vesta's gravity on September 5, 2012, and is scheduled to reach Ceres in February 2015.

Solar System Scoop

NASA has had ambitious plans to capture an asteroid for some time. The original plan was to send a spacecraft out to the asteroid belt to grab an asteroid and bring the captive back. But too many NASA programs have succumbed to budget cuts, and President Obama has challenged NASA to land an astronaut on an asteroid by 2025. The idea to capture a near-Earth asteroid arose as a less expensive alternative. To NASA administrator Charles Bolden, "The new plan is an ingenious alternative. If we can't get the asteroid, we'll wait for the asteroid to fly by us."

In addition to being the right size, the asteroid has to be traveling at a fairly slow speed (1.5 miles per second is the ideal) and maintaining an orbit that will bring it close to the Earth and the Moon in the early 2020s. In other words, the asteroid has to have the Right Stuff.

The captive will be pulled into orbit around the Moon, where astronauts can land on it and explore it. Beyond helping scientists to gain better understanding of asteroids, Bolden envisions the mission as an opportunity for developing technologies that would one day send people to Mars.

Meteors

The small objects composed of interplanetary dust or comet residue have three incarnations. A shooting star streaking across the sky is a meteor. Out in interplanetary space before it reaches Earth's atmosphere, it's a meteoroid. The occasional small (hopefully) rocks that fall to Earth are meteorites.

And when meteors shoot across the sky in droves, it's a meteor shower. The most prolific is the Perseid meteor shower, which peaks each year on August 12th as the Earth goes through dust left behind by comet 107P Swift-Tuttle on its 130-year orbit. On a clear night with no full Moon, the Perseids put on a spectacular show.

Figure 6. Perseid meteor shower.

Meteor Showers

During a meteor shower, the meteors seem to come from the same radiant or perspective point in the sky. The meteor showers are named for the constellation that appears to us as the radiant. In reality, the meteors are not coming at us from the constellation. What is happening is that the Earth is moving through the debris in that direction. Meteor showers seem to grow in strength after midnight, when our side of the Earth faces away from the Sun as the Earth goes through the dust.

The meteors we see streaking across the sky in a meteor shower are primarily minute dust particles burning up about 60 miles (100 kilometers) above the Earth. But meteors can be flashing by on any given night, at intervals of about every 10 minutes. These "sporadic" or "random" meteors tend be pebbles rather than dust.

Smart Facts

**Meteor Showers
Duration Number**

	Date of Maximum	Duration Above 25% of Maximum	Approximate Limits	Number Per Hour at Maximum	Parent Object
Quadrantids	January 4	1 day	Jan. 1-6	110	--
Lyrids	April 22	2 days	April 20-24	12	C/1861 G1
ε Aquarids	May 5	3 days	May 1-8	20	1P/Halley
δ Aquarids	July 27-28	7 days	July 15–Aug. 15	35	--
Perseids	August 12	5 days	July 25–Aug. 18	68	107P/Swift–Tuttle
Orionids	October 21	2 days	Oct. 16–26	30	1P/Halley
Taurids	November 8	Spread out	Oct. 20–Nov. 30	12	2P/Encke
Geminids	November 17	Spread out	Dec. 7-15	10	55P/Tempel–Tuttle
Taurids	December 14	3 days		58	3200 Phaethon

Usually, when meteors are streaking across the sky, the air is silent. Sometimes, although the light show is accompanied by the sound of them burning up. Some people compare the sound of a meteor shower to the faint sound of a train. Others describe them as "cracking" or "hissing" or even "swishing."

The trails usually linger for a few seconds after the flash. It is possible to reflect radio waves of the electrons in the trails. Beyond using them for scientific study, they can also be used to send messages.

Fireballs are meteors that are exceptionally bright. By IAU definition, a fireball is a meteor that glows brighter than any planet. These brightest of meteors not only outdazzle Venus, they are bright enough to cast shadows.

Every now and then, we are treated to an unusually active meteor shower: a meteor storm. The Perseids tend to be relatively predictable. On the other hand, the Leonids have years when their activity peaks. Every 33 years, they produce a meteor storm, with a pretty good storm the year before and after the peak show.

Abraham Lincoln was impressed by a meteor storm in 1833. The last Leonid peak was in 1999, when over 1,000 meteors went streaking by in one hour (Figure 7). For the next one, we have to wait until 2032; 2031 and 2030 should also be pretty spectacular, but that doesn't shorten the wait by much.

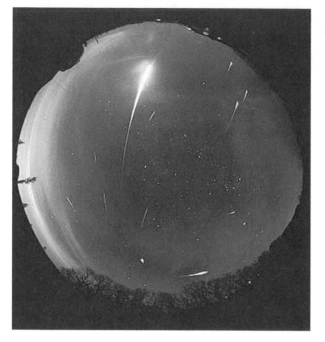

Figure 7

Meteorites

Meteoroid dust burns up completely in Earth's atmosphere. But larger pieces of space debris can survive the heat to land somewhere on Earth. These hardy objects are meteorites. In general terms, a meteorite is a natural object that lands on a celestial body from somewhere in space. Meteorites have landed on Mars and the Moon. On Earth, some have been known to land on (or in) people's houses or cars. Obviously, these events gain a lot of attention. In reality, they are very rare. People frequently find pieces of rocks they are convinced are meteorites and excitedly have them identified. The results are usually very prosaic—the presumed meteorites are ordinary rocks we can find on the ground every day.

Most of the meteors that land on Earth are stony, basically rocks. Others are mostly metallic, composed of an iron-nickel compound. Still others are classified as stony-metallic for their combination of both properties. Once on the ground, the iron-nickel meteorites are the easiest to spot as extraterrestrial objects, since they look different from our everyday rocks.

Antarctica is the best place for finding meteorites. Teams of researchers ride around the icy landscape on snowmobiles. Dark meteorites easily show up against the white snow. They could have fallen at any time and stayed buried under the ice for millennia until finally they were brought to the surface by ice movements and exposed by high speed antarctic winds..

Many meteorites match the minerals found in lunar rocks, confirming that they came from the Moon. A few of the meteors found on Earth have traveled from Mars. For years, the prevailing belief was that objects coming from Mars would burn up before they ever reached Earth. That has since changed dramatically with evidence that at least some have survived the trip. If ejected from Mars at relatively low speed, there are ways they could make it through interplanetary space and through the Earth's atmosphere without burning up.

The Mars meteorites are distinguished by their relative youth: part of the evidence that they really do come from Mars. Almost all the meteors found on Earth date back 4.5 billion years, the early years of the solar system. On the other hand, the small groups of meteorites from Mars are only 1.5 billion years old.

In addition to Mars and the Moon, a few meteorites come from Vesta. We know something about Vesta from the material, and now we have Dawn's data, still being analyzed.

Meteor Craters

The Earth is dotted with a few large meteor impact craters. Whether a collision with a meteoroid 65 billion years ago killed the dinosaurs (among other animal and plant species) and ended the Cretaceous period is up for dispute. However, there is compelling evidence of a massive impact. A thin layer of iridium is found throughout the world. Iridium is a rare metal more abundant in meteorites than in anything natural to Earth. The distribution of iridium on such a wide scale is consistent with such a powerful impact.

Solar System Scoop

Located in Arizona, the Barringer Meteor Crater is a mile-wide crater gouged roughly 50,000 years ago by meteor possibly 150 feet (50 meters) across. A hundred years ago, the surrounding land was dotted with bits of iron from the original meteoroid. It was also a time when many people thought they could get rich from mining. Many did. Daniel Barringer was not one of them. His mining scheme was not a financial success, but his idea that the crater was caused by a meteor impact is broadly accepted.

The discovery of the Barringer crater also gave weight to the theory that the craters on the lunar surface were caused by meteoroid impacts instead of volcanic activity.

Eugene Shoemaker analyzed the Barringer Crater, along with craters made by nuclear bombs on test ranges in the Southwest. His results showed a similar shape, confirming that the Barringer Crater resulted from an explosion. A meteoroid hitting the Earth would create that type of a blast.

The iridium layer would have been caused by a meteoroid six miles (10 kilometers) in diameter crashing to Earth. Fortunately, an event of that type is extremely rare, only happening once in tens of millions of years. Smaller objects survive the journey to Earth more often and can still have devastating effects.

The chance that an object with a diameter of at least 100 feet (91 meters) would come hurtling to Earth in the next 100 years has been estimated at 1%. NASA Administrator Bolden sees lassoing the asteroid and moving its orbit as in important first step in seeing that we are not hit by near-Earth objects of any substantial size.

CHAPTER 23

Exoplanets: Planets Beyond Our Solar System

In This Chapter

➤ Many exciting prospects, none prove to be real

➤ Pulsar planets: finally a real find

➤ Finding exoplanets through their gravity

➤ Finding exoplanets through transits

➤ COROT and Kepler and the search for Earth-like planets

Once the heliocentric view of the universe was accepted, the study of planets concentrated on celestial bodies revolving around the Sun. A variety of different objects were mistaken for planets over the centuries, but they still shared that one common characteristic. Astronomers have come up with reasonable models of how the planets in our solar system evolved eons ago. Theoretically, similar processes should have created planets revolving around other stars. Practically, there was no way of detecting them until the late twentieth century.

Isaac Newton recognized that other stars than our Sun could be harboring planets. In his essay "General Scholium," appended to the *Principia*, Newton wrote, "And if the fixed stars are the centers of similar systems, they will all be constructed according to a similar design and subject to the dominion of *One*." In the 1800s and into the mid-1900s, there were a number of claims of planets orbiting other stars, but all were ultimately dismissed.

Measuring star positions was a popular strategy in the search for new planets—and often led to erroneous claims. A notable example involved Barnard's Star, a star nearby the Earth. During the 1950s and 1960s, Peter van de Kamp of Swarthmore College claimed to have detected planets orbiting the star. As exciting as that prospect was, the observations supporting it proved false. The search for new planets was promising but thus far it was futile.

Pulsar Planets Discovered

In the 1990s, pulsars took the spotlight in the planet search. A pulsar is a peculiar, dead star that has used up all its internal energy and gone through a growing stage and then a collapsing stage. These "neutron stars" are very small and very speedy. They measure only 10 to 20 miles (16–32 kilometers) in diameter, but rotate in roughly one second. Some emit beams of radio waves, and some are positioned so that the beam crosses the Earth with each pulsar rotation. This produces radio pulses at very regular intervals, ranging from a few thousandths of a second to a few seconds.

Pulsar pulses are so extremely consistent that when a few pulsars seemed to deviate from the usual pattern, they drew intense scrutiny. In 1991, Andrew Lyne proclaimed to have found the first pulsar planet. It proved to be false, but shortly after came the first-ever claim that held up. Aleksander Wolszczan and Dale Frail discovered the first pulsar planet. Not only was their discovery supported by careful analyses, but there turned out to be at least two planets orbiting pulsar PSR B1257+12. The breakthrough had finally happened! In 1992, planets were finally found to be orbiting another star.

The brilliant discovery though, came with a catch. There is no way to visually see pulsar planets. And an object subject to continuous powerful bursts of radiation is very far from our idea of a Goldilocks planet. The Holy Grail in the search for new planets is finding planets like Earth, or at least capable of supporting life as we understand it.

The Search Goes On

In 1995, Swiss astronomers Michel Mayor and Didier Queloz announced that they used optical observations to find a planet. At the time, the idea was revolutionary. They used novel techniques to analyze spectra for signs that the frequencies marked in a star's spectrum were varying slightly but periodically. In accordance with Newton's laws, an object doesn't just move around by itself. A star that moves back and forth has to have an opposite force acting on it.

By meticulously measuring the star's movement toward and away from us, using the changing frequencies in what is known as the Doppler effect, Mayor and Queloz discovered an object that was not massive enough to be a star and not big enough to be visible. The logical choice was a planet.

Solar System Scoop

As a planet revolving around a star other than our Sun, the new discovery became known as an extra-solar planet, with "extra" signifying "outside" or "beyond." The unwieldy name is usually shortened to "exoplanet."

In the U.S., Geoff Marcy at San Francisco State University (now at the University of California at Berkeley) had been applying the same technique as Mayor and Queloz used in a series of measurements aimed at discovering more exoplanets. Marcy theorized that his team's first discoveries would be giant planets like Jupiter—or even bigger—and that they would take five years or more to orbit their stars (another Jovian property). Mayor and Queloz's exoplanet discovery made the journey around its star, 51 Pegasi (51 Peg, for short) in mere weeks. In other words, it had to be very massive to allow it to be detected—and very, very close to its parent star.

After first confirming the work of the Swiss scientists, Marcy and his team embarked on their own planet search. They have since discovered more than a hundred exoplanets.

As of June 2013, we know of 891 exoplanets. These objects occupy 695 planetary systems, including 133 multiple planetary systems. The total includes a substantial number of Earth-like planets, which we will get to later.

Given the tally of exoplanets, 28 represent just a small fraction of the growing collection. The 28 exoplanets in Figure 1 are notable in that their discovery was announced simultaneously. They are plotted in order of their distance from their parent star based on the radius of the Earth's orbit, 1 AU. Note that many of the planets are much closer to their host star than we are to the Sun.

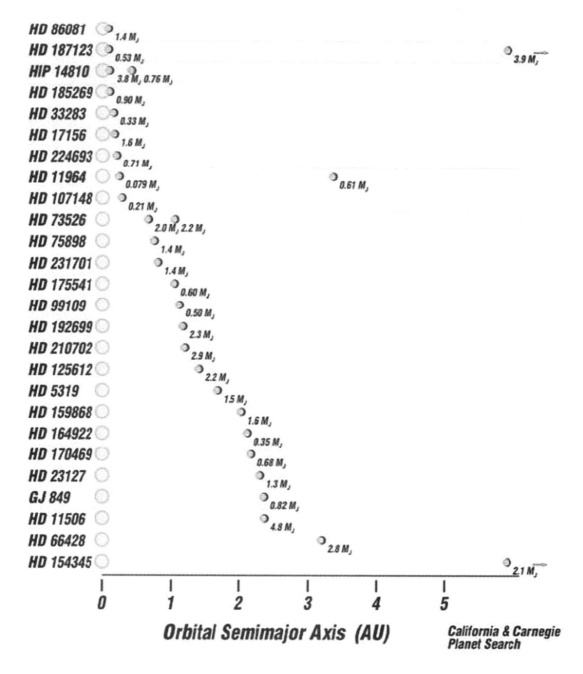

28 New Exoplanets

HD 86081 — 1.4 M_J
HD 187123 — 0.53 M_J — 3.9 M_J
HIP 14810 — 3.8 M_J, 0.76 M_J
HD 185269 — 0.90 M_J
HD 33283 — 0.33 M_J
HD 17156 — 1.6 M_J
HD 224693 — 0.71 M_J
HD 11964 — 0.079 M_J — 0.61 M_J
HD 107148 — 0.21 M_J
HD 73526 — 2.0 M_J, 2.2 M_J
HD 75898 — 1.4 M_J
HD 231701 — 1.4 M_J
HD 175541 — 0.60 M_J
HD 99109 — 0.50 M_J
HD 192699 — 2.3 M_J
HD 210702 — 2.9 M_J
HD 125612 — 2.2 M_J
HD 5319 — 1.5 M_J
HD 159868 — 1.6 M_J
HD 164922 — 0.35 M_J
HD 170469 — 0.68 M_J
HD 23127 — 1.3 M_J
GJ 849 — 0.82 M_J
HD 11506 — 4.8 M_J
HD 66428 — 2.8 M_J
HD 154345 — 2.1 M_J

Orbital Semimajor Axis (AU)

0   1   2   3   4   5

California & Carnegie
Planet Search

Figure 1

Marcy was right in claiming that the hunt for new planets would turn up planets more massive than Jupiter. Many exoplanets make giant Jupiter look downright puny. He was less accurate in predicting they would have a long year. In fact, the most surprising fact about exoplanets is that so many are very close to their stars and complete their orbit in just a few days. This unusual combination has earned the new planets the name "hot Jupiters": Jupiters because they are hugely massive and must be gas giants, and "hot" because of their close proximity to their stars.

The preponderance of massive planets among the known exoplanets has a practical explanation: massive planets are easier to detect by virtually all techniques. Among the countless planets yet undiscovered, less massive planets may actually outnumber the giants. The hunt is on to find more of the less massive ones. So far, all but 50 of the known exoplanets have more than 10 times the mass of our Earth.

Theoretical models have long explained why our solar system is structured the way it is, with low-mass rocky planets closer in to the Sun and gas giants farther out. Powerful solar winds from the nascent Sun would have blown much of the hydrogen and helium from the inner planets, without stripping them from the atmospheres of those planets farther out in space. Now that we know there are other solar systems configured differently some ideas had to be changed. The predominant theory is still that giant planets evolve at least 5 AU from their parent stars, but there is some dynamic that pushes them inward so they are eventually close to those stars. Long-held ideas are continually changing with new discoveries.

Imaging Exoplanets

In practical terms, looking for an exoplanet next to its star means trying to find a faint object next to a hugely bright one: the brilliance totally overpowers and obscures it. Astronomers have confronted this problem since the mid-1800s, when a faint white dwarf star was discovered next to Sirius, the brightest star in the sky. The contrast between an exoplanet and its parent star is even more extreme. On the positive side, our technology is far superior, and several research teams have been testing and improving techniques for suppressing a bright object in order to see a faint object next to it.

In 2008, a team of astronomers detected a group of three exoplanets orbiting the star HR 8799 (Figure 2). They applied innovative techniques to both data gathering and image processing. They allowed the field of view to rotate as they captured the images so that the actual exoplanets were distinct from the artifacts caused by the optics. They also used adaptive optics on the Keck and Gemini telescopes in Hawaii. In this technique, which we discussed earlier, plungers push the back of the secondary mirror, thereby counteracting atmospheric distortion.

The team's detailed analysis confirmed that the objects they observed were indeed planets. It was the first observation of a planet around a star other than our own Sun.

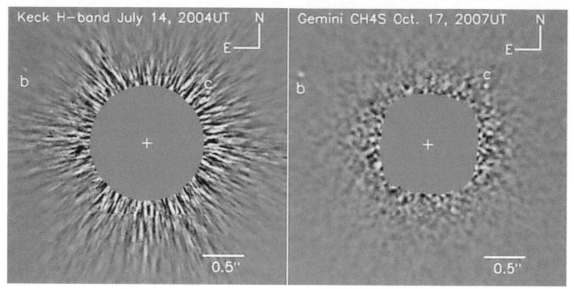

Figure 2

As an alternative to suppressing the glaring light of the close-up star, astronomers have come up with another way to make exoplanets easier to observe: detecting close faint stars using NASA's Galaxy Evolution Explorer satellite. Being next to a young, faint star makes the planet easier to pick out. The catch is that dim stars are harder to find. But as an advantage, these volatile young stars give off more ultraviolet light than more mature stars. Galaxy Evolution Explorer is equipped with highly sensitive ultraviolet detecting equipment.

The targets are low-mass M-Class stars, or red dwarves. They have three properties that make them ideal for seeking out planets: they are relatively close to Earth and in a clear field of vision, they are fairly faint, and they sizzle in ultraviolet. Finding young faint stars may prove easier than blocking the glaring light of their older counterparts.

NASA's Exoplanet Exploration Program

Known as ExEP, NASA's Exoplanet Exploration Program leads the quest for Earth-like planets. The first phase of the project involves exploring and characterizing the numerous types of planetary systems populating the universe. That basic understanding paves the way for seeking out planets that might be habitable.

The Terrestrial Planet Finder (TPF) is equipped with a coronagraph to shade the star's glaring light to reveal Earth-like planets orbiting in the habitable zone. (The habitable zone, approximately 1 AU for a star comparable to the Sun, is the area where the star's energy creates conditions on the planet's surface reasonably conducive to the presence of liquid

water.) An alternative strategy involves separated spacecraft and telescopes directing beams of light to a central spacecraft that acts as an inferometer to hide the star's light and pick up the heat radiated by Earth-like planets.

Measuring Star Positions from Earth and from Space

Astrometry, the measurement of star positions, has been a basically unsuccessful technique for finding new planets from Earth. In 2009, it appeared that ground-based scientists had finally succeeded in finding a planet using very accurate, detailed measurements of the position of the small star VB 10 (Figure 3).

Solar System Scoop

In January 2012, a team led by astronomers at NASA's Exoplanet Science Institute at the California Institute of Technology used data from Kepler to discover the three smallest exoplanets found yet. The planets orbit the star KOI-961. The smallest of the trio is roughly the size of Mars. All three planets appear to be rocky like the smaller planets in our own solar system. Very few known exoplanets are terrestrial. Unfortunately, this small, rocky family orbits too close to the parent planet to be in the habitable zone.

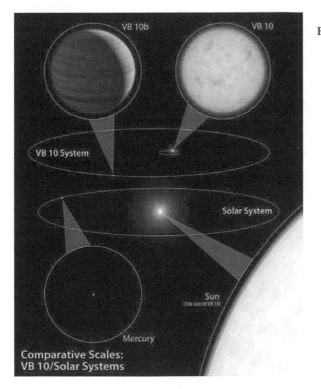

Figure 3

VB 10 is one of the faintest of the faint red dwarves. It is also very close to the Earth. Astronomer George Van Biesebroeck discovered it in 1944 near a much brighter companion star. So close to the Earth, it can be seen moving across the sky at a very fast rate (for a star): one second of arc per year. It has a very high "proper motion," that is, movement across the sky as opposed towards us or away from us along the line-of-sight, as gauged by the Doppler technique used to discover the first exoplanets.

When measurements of VB led to announcements of a new planet, it was a credible claim. The presumed planet was reported to have a mass of at least six Jupiters and an orbit of 272 days and 0.36 AU from the star. The juxtaposition of a star with a mass 10% of the Sun's and a planet with six times Jupiter's mass could mean that finally, a planet had been found with astrometry.

Further observation with sensitive radial-velocity measurements deflated the celebrated new discovery. No planet could be found.

Astrometry had once again failed to detect a new planet from the ground. It has proved more useful from space. The Hubble Space Telescope is equipped with the Fine Guidance Sensors, which help it lock onto faint stars. Hubble has captured the motion of planets that had been discovered by radial velocity. It also found by tracking an object's orbit that it was much more massive than initially thought. In that case, the purported planet turned out to be a brown dwarf.

Astrometry can be useful in the search for planets after all, but best left to the Hubble Space Telescope.

Discovering Planets by Transit

We discussed transits in our own solar system, like the rare transits of Venus. When Venus transits the Sun, it blocks roughly 0.1% (1/1,000th) of the sunlight for about six hours. The dip in the sun's light is easily gauged. Over the last few years, transits of exoplanets have been found by monitoring numerous stars. With technology such as electronic cameras known as CCDs (charge-coupled devices), amateur astronomers can join professionals in observing exoplanet transits.

COROT

In 2007, the French satellite COROT (**Co**nvection, **Ro**tation, and Planetary **T**ransits) was launched into polar Earth orbit. The "T" part of its mission involves monitoring some 120,000 stars by looking at 12,000 fields of view, each inhabited by about 100 stars. Though the scientists surmised that most exoplanet discoveries would be massive giants, they are especially seeking out smaller terrestrial planets. Their strategy is to observe the same fields of view repeatedly for 150 days, so that planets with orbits of less than 50 days would experience three transits. A planet with such a quick orbit would be very hot unless its host star was dim. And revolving around a cool star, the planet might be temperate enough to support water and life.

Its first year in operation COROT discovered two planets, and two more were found in 2008. In 2009, COROT detected what was then the smallest exoplanet discovered. Located in the constellation Monoceros 500 light years away, COROT 7b has a diameter only 1.7 times the size of the Earth and a mass 4.8 times Earth's mass. From its mass and its size, we know that its density is similar to Earth's. Only its orbit diverges dramatically from Earth's. At 20 times closer to its star than Mercury is to the Sun, COROT 7b speeds around in 20.5 hours: less than one Earth day.

The discovery of a small rocky planet similar to Earth was especially exciting. Then in March 2010, COROT 9b was reported, with 80% the mass of Jupiter and an orbit similar to Mercury's. It was the first transiting temperate planet similar to the denizens of our own solar system.

By the end of 2011, the tally of COROT's exoplanet discoveries was up to 24, with several hundred additional candidates waiting for confirmation.

Kepler

As part of ExEP, NASA's Kepler mission is an even bigger venture to discover exoplanets by transit. Since 2009, Kepler has been monitoring the brightness of more than 145,000 stars in one field of view encompassing large areas of the constellations Cygnus and Lyra. It changes fields of view every few months.

The first discoveries, announced in 2010, included several planets close to Jupiter's mass. Kepler is well-equipped to discover smaller, terrestrial planets like Earth as illustrated by the following observation: After a star's light dips when its planet moves in front of it, blocking a speck of bright starlight, there is an even tinier dip as much fainter planet moves behind the star. Kepler showed it was sensitive enough to detect that miniscule dip in the total light caused by the planet going behind the star.

Figure 4. An artist's conception of NASA's Kepler Mission, which was launched in 2009 to search for exoplanets.

Kepler proved its capabilities in practical terms when it found the trio of small terrestrial planets. As of February 2013, Kepler has discovered 114 confirmed exoplanets in 69 star systems, out of a candidate pool of 2,740 possible exoplanets. Its mission has been extended to 2016.

Examining Exoplanets

When a planet moves in front of its star, any atmosphere it might have distorts a bit of the star's light (Figure 5). Comparing the spectrum of the merged planet-star system when the planet is in transit with the spectrum when the planet is behind the star or off to the side (and thus invisible), reveals properties of the planet's atmosphere.

That technique has allowed traces of water and other molecules to be detected in a few exoplanet atmospheres. For example, when its planet transits the star HD 189733b in the constellation Vulpecula, it dims the total light by roughly 3%.

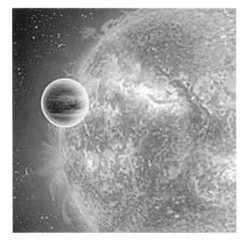

Figure 5

A few planets have been discovered through microlensing, observing the way their gravity focuses light from stars that they pass in front of. Unfortunately, it's impossible to learn much about a planet from a single event, and such events are extremely rare.

In a January 2013 study, astronomers at Caltech concluded that the Milky Way Galaxy is home to at least as many planets as stars: that makes a staggering 100 billion to 400 billion exoplanets awaiting discovery.

The Sun

Our Sun: Shining at the Center

After centuries of domination by the geocentric view of the universe, our Sun was finally given its rightful position as the center of our solar system. It may have lost some of its unique status with the discovery that there are countless stars also harboring planets revolving around them. (Though the fact that they are formally called extra-*solar* planets maintains our Sun as the central reference point.) Nonetheless, the Sun is essential to life as we know it on Earth. And its gravity governs the orbits of all the myriad objects—from gas giants to flashing bits of debris—that populate our very interesting realm of space.

The Sun, like all other stars, is made up mainly of hydrogen (90%). Helium accounts for about 9% and all the other elements add a scant 1% to the total. Of all the solar system planets, Jupiter and Saturn are closest in composition to the Sun: almost completely hydrogen and helium. On our own planet and its terrestrial companions, most of the hydrogen and helium escaped eons ago: the result of our closer proximity to the powerful Sun.

Sunspots

Once Galileo aimed his telescope at the sky, he spent countless hours peering at every object he could pick out with his primitive lens (even if he was sometimes wrong about what it was). The Sun is too bright for telescope viewing, but the device was still useful for projecting an image onto a wall or a screen. One day in 1610, something that looked like blemishes appeared on the sun's disk. Galileo was looking at sunspots.

Galileo was not alone in his discovery. At least two other observers independently saw the same phenomenon and inevitably, a controversy ensued. Indeed, the dispute over who deserved credit might have played some role in Galileo's fight with the Catholic Church. Sunspots also challenged Aristotle's idea that the Sun was fixed and unchanging.

Historically, neither Galileo nor his peers were the first to observe sunspots. The earliest recorded descriptions date back to antiquity. By A.D. 28, Chinese astronomers were keeping track of sunspot activity. In the West, scholars in the 9th and 13th centuries described sunspots but thought they were transits of planets. Galileo's explanation was accurate.

Galileo published a book on sunspots in 1613. The work included a series of engravings depicting the phenomenon: 17th-century animation. The drawings illustrate the way the Sun rotates and the individual sunspots and sunspot clusters change and evolve. Galileo knew that the Sun has a rotation of about one month. He clearly understood that the Sun was *not* a stationary object: Copernicus was right and Aristotle was wrong.

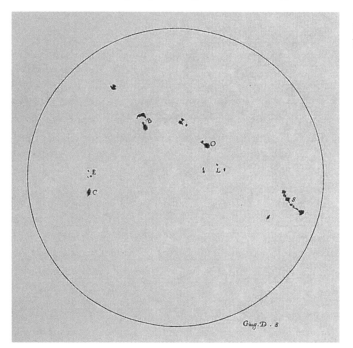

Figure 1. A sunspot drawing by Galileo

Sunspots appear in clusters, in some cases with dozens dotting the Sun at the same time. An individual sunspot has a dark center, or *umbra* (Latin for shadow), and a lighter filamentary area outside called the penumbra.

In the late 17th century and into the first half of the 18th, sunspots rarely appeared. When the odd sunspot showed up, it was so unusual that it immediately attracted attention and was written up as something scientifically notable. These references are compelling evidence that there was an actual dearth of sunspots as opposed to a lack of interest or understanding.

Solar System Scoop

At the turn of the 20th century, astronomer Walter Maunder wrote papers on changing sunspot latitudes in which he documented the period of low sunspot activity. His work was rediscovered in the 1970s by Jack Eddy, who titled his landmark article on the topic "The Maunder Minimum." The name stuck (though it is also known as prolonged sunspot minimum). It refers to the period from 1645 to 1717.

In Europe, the Maunder Minimum coincided with a period of unusually cool temperatures known as the Little Ice Age. Whether those cool temperatures were global or were actually connected with the low sunspot activity is unknown. It would be useful to know, as knowledge of any relationship might further our understanding of the current global climate change.

By the mid-19th century, scientists were aware that the number of sunspots increased and decreased in cycles averaging 11 years. In some cycles, the number of sunspots and sunspot groups is higher at maximum, which could be a sign of some longer term cycle.

In a tradition dating back to the 19th century, astronomers usually keep track of sunspots in terms of "sunspot number." Sunspot number is not the actual number of sunspots; rather, it is 10 times the number of sunspot groups plus the total number of individual sunspots. This system of counting adds significance to the grouping of sunspots. There should be an equal number of sunspots and groups on the Sun's near and far sides (though we have to conjecture about the far side).

The last peak of the sunspot cycle occurred in about 2004–5. Given that, we expected to see a sunspot minimum in about 2008–9, with an increase following. However, the minimum fell particularly low: for roughly 75% of the days of 2008 and 2009 no sunspots were seen at all, equaling sunspot number 0. The minimum lasted an unusually long time. In fact, it was the longest minimum and the lowest sunspot count in the last hundred years. (No sign of a Little Ice Age!)

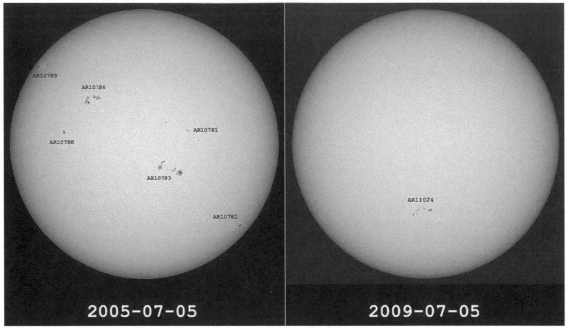

Figure 2. Solar cycle mazimum in 2005 and minimum in 2009.

Sunspots are magnetic, which is the key to knowing when they start. This property of sunspots was discovered in the early 20th century by George Ellery Hale. Through his work at the Mount Wilson Observatory in Pasadena, he discovered that sunspots have magnetic fields 3,000 times the magnetic field of the Earth's average magnetic field.

As it works, for each 11-year period, the leading spot in each pair of spots in the Sun's northern hemisphere would have one of the magnetic polarities; say north. The following spot in that pair would have the opposite polarity, or south. In the southern hemisphere the polarities would be reversed: south first, north second.

At sunspot minimum, the sunspots disappear and the magnetic field is completely dispersed around the Sun. The sunspots of the new cycle, with the polarities of the leading and following spots reversed, begin to appear at mid-latitudes, roughly 40 degrees north and south of the equator. (This actually makes the true sunspot cycle a magnetic cycle of 22 years.) The sunspot clusters pop up closer to the equator as the cycle heads toward the maximum. At the peak, the groups extend to roughly 10 degrees north and south of the equator: the closest to the equator they get.

The next sunspot peak should be in 2015. Some predictions place it earlier; it may even appear by the end of 2013.

A Typical Day at the Sun

Barring haze, clouds, or solar eclipse, the Sun brilliantly lights up our day. The massive object gets its energy from nuclear fusion deep in its interior. At extraordinarily high density and a scorching temperature of 30 million degrees F (15 million degrees C), the Sun is busy fusing hydrogen into helium, converting particles into energy in the process. The amount turned into energy follows Einstein's famous formula: E = mc2. It keeps the Sun going for its 10-billion-year lifetime. So far, our star still has half its life left.

Energy from the solar core radiates upward, beginning as gamma rays. It takes roughly a million years for the gamma rays to project upward. In the outer third or so, energy is carried upward by a boiling process known as convection. In high-resolution images of the Sun, those convection zones, called granules, appear as areas about the size of Texas.

The Sun embodies the term "gas giant": it is entirely gaseous with no solid or liquid anywhere. Each particle of gas is opaque to some degree. This property keeps us from looking farther into the Sun the same way that fog limits how far we can see down the road.

The level at which we can no longer see into the sun is the photosphere (*photo* meaning light in Greek). When we think of the Sun, what we usually think of is the solar photosphere: the source of the light that illuminates our planet and keeps us warm.

The sunspot images in Figure 2 show the solar photosphere. If you look at the photosphere near its edge (or limb), you may notice evidence of *limb-darkening*. The darker appearance comes from the fact that by looking near the limb of a spherical object like the Sun we are looking diagonally through the gas. The opacity builds up faster and blocks our view at a higher level. The temperature there is roughly 11,000 degrees F (6,000 degrees C).

Put together the facts that the Sun appears darker near its edge and that cooler gas sends out less powerful rays, and we can infer that the level we see is cooler higher up. Thus the darkening of the solar limb means that the temperature of the photosphere increases with altitude.

We receive light from the solar atmosphere by means of a rather odd process. Almost all the light we get from the Sun in the visible and the infrared comes from the negative hydrogen ion. Normal hydrogen has one proton, with one electron circling around it. An ordinary ion (also called a positive ion for its positive charge) is only a proton. Hydrogen, however, has an unusual property. It can be in a state where the atom can hold a second electron, though very weakly and very briefly. That makes it a negative ion.

Releasing the second electron sends energy out into space. The electron can fly out at any velocity, so the change in energy covers the full spectrum.

Eclipses, the Chromosphere, the Photosphere, and the Corona

The photosphere is the lowest level of the solar atmosphere. The higher levels are normally too faint to see—until we get a solar eclipse.

Approximately every 18 months, the Moon's orbit around the Earth (inclined 5 degrees) intersects with the Earth's orbit around the Sun at a point when the three celestial bodies are precisely aligned. Beyond this perfect alignment, the Moon must be at a point where it is fairly close to the Earth and so appears larger than usual. When these things come together, we get a total solar eclipse (more about eclipses later in this chapter).

During a total solar eclipse, the Moon covers the photosphere. In some cases, you can blink and you miss the event. Other times, the Sun may be hidden from sight for more than seven minutes.

When the Moon first moves in front of the Sun, a red rim appears. This reddish area of the Sun is the chromosphere, from the Greek word *chromo* for "color."

Smart Facts

Solar prominences rising up off the Sun are the same reddish color as the chromosphere. Many remain out of our view for weeks at a time, while others rise or change in mere hours.

Prominences should not be confused with solar flares, which are much more volatile. People sometimes erroneously refer to "flares on the edge of the Sun" in images of an eclipse. Those "flares" are really prominences.

The chromosphere and the prominences share their reddish color because most of their radiation comes from specific colors given off by hydrogen gas. Hydrogen's red emission is the most powerful one.

The chromosphere is composed completely of tiny dynamic jets called "spicules" (little spikes) that are continually rising and falling. From a temperature minimum near the top of the photosphere, spicules rise up by a factor of about two, to roughly 22,000 degrees F (12,000 degrees C) or even hotter. Hundreds of thousands of them dot the solar surface at a given moment, rising and falling in about 15 minutes.

As the Moon's black silhouette is about to hide the solar photosphere, the chromosphere comes into view. Given the craggy lunar terrain with mountains and valleys, some valleys are always deeply embedded in the Moon's silhouette. As a result, a few beams of sunlight are still stubbornly visible as totality begins, sneaking through the lunar valleys to our eyes. Known as Baily's beads, because when they are visible, they appear as a string of dots along the lunar limb, from an 18th-century observation, they gradually disappear.

The last Baily's bead gleams so brilliantly on the edge of the Sun and the Moon's silhouette that it looks like a diamond engagement ring. As a result, it's known as the diamond ring effect (Figure 3).

Figure 3

Until the very last seconds of the diamond ring effect, too much of the photosphere is visible to make it safe to stare at the Sun. But from that point though the time of total coverage until the next diamond ring appears as totality ends, it is completely safe to look up at the Sun.

What we see then is truly amazing: the solar corona or "crown of light" emerges ("corona" comes from the Latin word for crown), surrounding the darkened Sun.

The blazing gas in the corona is kept from escaping by the Sun's magnetic field. Its temperatures soar to at least 3 million degrees F (1 million degrees C).

The corona's shape changes over the 11-year sunspot cycle since its magnetic field fluctuates with the magnetic field coming from sunspots. When the sunspot cycle is near maximum ("solar maximum"), there are sunspot groups even at very high latitudes and the streamers of hot gas held in place by the coronal magnetic field jet out in all directions. At that point, the corona is pretty much round. Closer to solar minimum, there are just a few sunspot groups, and they cluster at lower latitudes. As a result, the solar corona has relatively few streamers, and they are mostly concentrated near the equator (Figure 4). The corona then is less elongated.

Figure 4

The question of how the corona gets heated to such extraordinarily high temperatures has fueled a lot of intensive research. Despite that, there is still no definitive answer. Ordinary radiation couldn't do it, since it involves a hotter gas on top of a cooler gas. Gas would be expected to be hotter, not cooler as it moves away from the Sun.

According to one set of theories, the magnetic field might be the key. Waves in the magnetic field might be heating up the corona. Research teams have been attempting to detect such waves through observations with ground-based solar telescopes.

An alternative set of theories focuses on nanoflares, tiny, ongoing explosions in the solar photospheres that could be heating up the corona.

Knowledge of the dynamics causing the blazing heat in the corona has significance beyond our own Sun. Our galaxy is populated by billions of similar stars and there are millions of other galaxies in the universe with their own stellar inhabitants. Understanding our Sun's idiosyncrasies will help us to understand other stars with similar chromospheres and coronas. Some of them may be harboring Earth-like planets.

Total Eclipse of the Sun

A series of total eclipses occur over a period of 18 years, 11.5 days. Eclipses of all durations happen within that time frame. The alignment of the Earth, Moon and Sun that caused an eclipse lasting just over seven minutes in Africa in 1973 was repeated in Mexico in 1991, this time lasting almost seven minutes, and again in the Pacific Ocean off the Coast of China in 2009, falling just short of seven minutes.

A solar eclipse happens when the lunar shadow hits the Earth. With the Sun so much bigger, the Moon's shadow tapers, forming a conical shape. The Earth intercepts the tapering shadow somewhere near the cone's tip. About half the time, the Earth crosses the shadow before the cone's tip, creating a total solar eclipse (Figure 5).

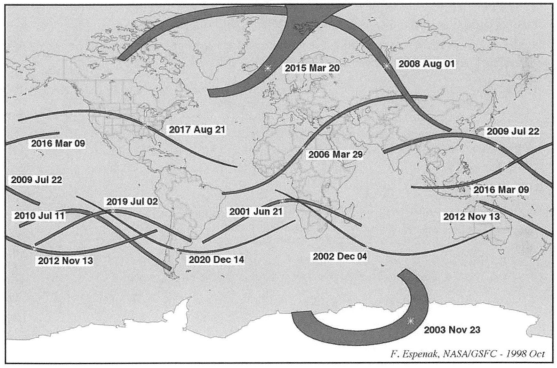

Total Solar Eclipses: 2001 - 2020

Figure 5

As the Earth rotates and the Earth and the Moon continue along their orbits, the Moon's shadow sweeps out a region extending thousands of miles in length, but only about 200 miles (350 kilometers) in width, or even narrower. It can be wider when it surrounds the Earth near the poles. Eclipses are longer when the path approaches the equator near noon, when the Earth's rotation keeps up with the speed of the lunar shadow to the greatest degree.

In 1973, an eclipse tracked the equator near noon and slowed up as a result of the Earth's rotation just enough that a supersonic Concorde kept up with totality for 74 minutes. Telescopes in special windows on top of the plane were aimed at the Sun high overhead.

On the whole, few people get to see an eclipse for that long, even from airplanes. Eclipse tourism, however, has grown exponentially as an offshoot of ecotourism as well as amateur astronomy. People travel from all parts of the world to the band of totality to marvel at all aspects of the awesome event.

The long eclipse of July 22, 2009, was followed by a relatively long eclipse on July 11, 2010. Totality was only visible at Easter Island in the Pacific and some atolls near Tahiti, but that did not dissuade observers who traveled to those locations or booked tickets on cruise ships and airplanes that deliberately traveled that path.

The total eclipse of November 12, 2012, began in the sky over northeastern Australia and was visible in the Australia-Pacific region and some parts of South America.

The next total eclipse will take place on November 3, 2013, and be visible in Africa, crossing Gabon, Congo, and Kenya. After that, a total eclipse will cross the Arctic on March 20, 2015. It will cross over Svalbard, the home of the armored bears in *The Golden Compass*.

For a total eclipse visible in the U.S., we'll have to wait for August 21, 2017. It will begin over the border of California and Oregon, traversing the country diagonally to South Carolina.

The last total solar eclipse of this decade will occur on July 2, 2019. Perhaps fittingly, it will cross through Chile, home of some of the world's biggest telescopes. Six months later, on December 14, 2020, another total eclipse will cross Chile and Argentina.

Following that, the next eclipse visible over North America will be in 2024. It will be the first total eclipse to cross Canada since 1979.

For those who are not in the choice locations, a partial eclipse is visible from areas within about 500 miles (800 kilometers) from the path of the eclipse. Like partial viewing seats in a theatre, the farther off to the side, the more it compromises the performance. In the case of an eclipse, it means the smaller the chunk that gets taken out of the Sun. Baily's beads, the diamond ring, the chromosphere, and the corona are all absent from a partial eclipse. In addition, you'll need to look through a special filter or view a projected image to be safe from the Sun's glare. But seeing part of the Sun disappear may be a practical alternative to a long trip to the site of totality.

Annular Eclipses

Annular solar eclipses occur at roughly the same rate as total eclipses. During an annular eclipse, the tip of the lunar shadow, its umbra, falls just short of the Earth. Instead, we are in the path of the anti-umbral shadow. A ring or annulus of sunlight stays visible around the Moon, which is slightly farther from Earth than usual. The event takes its name from the annulus of visible sunlight. Colloquially, it's often called a "ring of fire" eclipse.

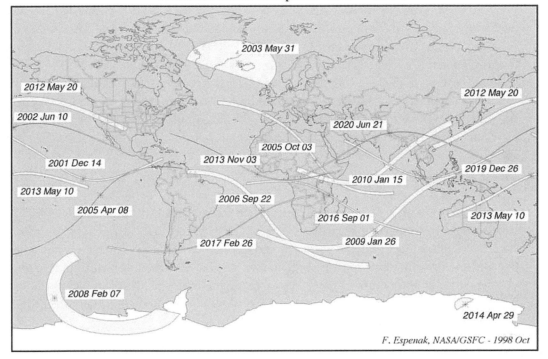

Annular Solar Eclipses: 2001 - 2020

F. Espenak, NASA/GSFC - 1998 Oct

Figure 6

Like a partial eclipse, an annular eclipse has to be viewed through a special solar filter. Cameras need to be fitted with similar solar filters.

The last annular eclipse awed observers in Australia on May 9–10, 2013. The next ring of fire event will be visible from areas in and near Antarctica on April 29, 2014 (with more penguin than human observers). After that, an annular eclipse will take place on September 1, 2016, in Gabon, Congo, and crossing Africa to Madagascar, and then on February 16, 2017, in Chile and Argentina, finally crossing the Atlantic to Angola.

The Green Flash

The green flash is an intriguing phenomenon caused by the effects of the Earth's atmosphere as we look at a setting Sun. For a few seconds, just as the Sun sets over a distant horizon, a green flicker can appear at the top of the Sun. The best viewing spot is a beach or any site looking out over an ocean.

The Earth's atmosphere distorts incoming sunlight by bending it downward. The light is bent by different amounts for different colors. A setting Sun is really the Sun a degree or two below the horizon with its image bent so we can still see it. At that point, there are two overlapping images of the Sun, with red the lowest and blue the highest. (To remember the colors of the spectrum, you can use the mnemonic ROY G BIV: red orange yellow green blue indigo violet.) When the Sun is very low, the shorter blue, indigo, and violet rays are scattered by the atmosphere (the shortest wavelengths are most easily scattered). Water vapor in the atmosphere causes the orange and yellow to dissipate. What is left then, are slightly overlapping images in red and green. When the upper red edge of the Sun sets, that leaves the green upper edge by itself—for a fleeting moment.

Stormy (Space) Weather

The Sun is essentially a ball of energy. A solar flare results when the tremendous energy stored in the magnetic field in sunspots is explosively ejected. When these solar explosions occur, the high-energy particles and x-ray and gamma-ray radiation reach our atmosphere in eight minutes (the same as ordinary light from the Earth from the Sun). When these volatile particles hit our atmosphere, they cause magnetic storms. Particles hitting the Van Allen Belt cause or magnify the auroras. All these Sun–Earth interactions fall under the heading of what is now called space weather.

We now have space weather forecasts, the same way we have our more prosaic terrestrial weather forecasts. Solar storms threaten to damage satellites in orbit around the Earth. They can certainly be dangerous to astronauts, and even to passengers in high-flying airplanes, especially those following solar routes. Even for those of us on terra firma, it is important to be able to predict space weather. Daily space weather forecasts are among the many useful and interesting resources available online.

The twin satellites of NASA'S STEREO (**S**olar **TE**rrestrial **RE**lations **O**bservatory), launched in 2006, lead and trail our planet as it orbits the Sun, providing details of the far side and tracking the flow of energy and particles from the Sun to the Earth. In 2010, NASA launched the Solar Dynamics Observatory, which transmits images of the Sun and any storms happening every 10 seconds.

But the most impressive solar explorer is comet Lovejoy, which accomplished something no artificial satellite could do. In 2011, the dirty snowball plunged into the Sun's fiery corona. Scientists using STEREO and the Solar Dynamics Observatory watched the comet's tail darting through the solar atmosphere and used the data to create a model. "Wiggles" in Lovejoy's tail seemed to be signs of complex interactions between charged particles in the comet's tail and the solar magnetic field.

Lovejoy is quite literally a trailblazer. Similar studies using comets as "natural solar explorers" should come in the future. NASA's Solar Probe Plus is planned for launch in 2018. For perspective, the probe will venture within about 3.7 miles of the blazing solar surface. Lovejoy came within 87,000 miles—more than 40 times closer.

Comets were not only maligned as bad omens, they were also called "vermin of the sky" by scientists frustrated by comets that interfered with their observations. Now their characteristics are harnessed to explore the Sun: the center of our solar system.

CHAPTER 25

Readings and Web References

Selected Readings

Non-Technical Magazines

Sky and Telescope, P.O. Box 9111, Belmont, MA 02138, 800 253 0245, www.skyandtelescope.com.

Astronomy, 21027 Crossroads Circle, P.O. Box 1612, Waukesha, WI 53187, 800 533 6644, www.astronomy.com.

Mercury, Astronomical Society of the Pacific, 390 Ashton Ave., San Francisco, CA 94112, 800 962 3412, www.astrosociety.org.

StarDate, 2609 University, Rm. A2100, University of Texas, Austin, TX 78712, 800 STARDATE, www.stardate.org.

The Griffith Observer, 2800 East Observatory Road, Los Angeles, CA 90027, www.griffithobs.org.

Science News, P.O. Box 1925, Marion, OH 43305, 800 552 4412, www.societyforscience.org.

Scientific American, P.O. Box 3186, Harlan, IA 51593-2377, 800 333 1199, www.scientificamerican.com.

National Geographic, P.O. Box 96583, Washington, DC, 20078-9973, 800 NGC LINE, www.nationalgeographic.com.

Natural History, P.O. Box 5000, Harlan, IA 51593-5000, 800 234 5252, www.naturalhistorymag.com.

New Scientist, 151 Wardour St., London W1V 4BN, U.K., 888 822 4342, rbi.subscriptions@rbi.co.uk; www.newscientist.com.

Physics Today, American Institute of Physics, 2 Huntingdon Quadrangle, Melville, NY 11747, 800 344 6902, www.aip.org/pt; www.physicstoday.org.

Science Year (World Book Encyclopedia, Inc., P.O. Box 11207, Des Moines, IA 50340-1207, 800 504 4425, The World Book Science Annual, www.worldbook.com.

Smithsonian, P.O. Box 420311, Palm Coast, FL 32142-0311, Washington, DC, 20560, 800 766 2149, www.smithsonianmag.com.

Discover, P.O. Box 42105, Palm Coast, FL 34142-0105, 800 829 9132, www. discovermagazine.com.

The Planetary Report, The Planetary Society, 65 North Catalina Avenue, Pasadena, CA 91106-2301, 818 793 5100, www.planetary.org.

Observing Reference Books

Jay M. Pasachoff, *A Field Guide to the Stars and Planets*, 4th ed. (Boston: Houghton Mifflin Co., 2000; updated 2003). All kinds of observing information, including monthly maps and the 2000.0 sky atlas by Wil Tirion, and Graphic Timetables to locate planets and special objects like clusters and galaxies.

Observer's Handbook (yearly), Royal Astronomical Society of Canada, 136 Dupont Street, Toronto, Ontario M5R 1V2 Canada.

Guy Ottewell, *Astronomical Calendar* (yearly) and *The Astronomical Companion* Department of Physics, Furman University, Greenville, SC 29613, 864 294 2208, www. skyandtelescope.com.

For Information About Amateur Societies

American Association of Variable Star Observers (AAVSO), 25 Birch St., Cambridge, MA 02138, www.aavso.org.

American Meteor Society, Dept. of Physics and Astronomy, State University of New York (SUNY), Geneseo, NY 14454, www.amsmeteors.org.

Astronomical League, the umbrella group of amateur societies. For their newsletter, *The Reflector*, write The Astronomical League, Executive Secretary, c/o Science Service Building, 1719 N St., N.W., Washington, DC, 20030, www.astroleague.org.

Astronomical Society of the Pacific, 390 Ashton Ave., San Francisco, CA 94112, www. astrosociety.org.

International Dark Sky Association, c/o David Crawford, 3545 N. Stewart, Tucson, AZ 85716, 877 600 5888. www.darksky.org.

The Planetary Society, 65 North Catalina Ave., Pasadena, CA 91106-2301, 818 793 5100, www.planetary.org.

British Astronomical Association, Burlington House, Piccadilly, London W1V 0NL, England, www.britastro.org.

Royal Astronomical Society of Canada, 124 Merton St., Toronto, Ontario M4S 2Z2, Canada, www.rasc.ca.

Saturn

NASA and The Hubble Heritage Team (STScI/AURA) • Hubble Space Telescope WFPC2 • STScI-PRC01-15

These images of Saturn, taken with the Wide Field Planetary Camera 2 onboard Hubble, were collected by Richard French (Wellesley College), Jeff Cuzzi (NASA/Ames), Luke Dones (SwRI), and Jack Lissauer (NASA/Ames), and have been prepared for presentation by the Hubble Heritage Team.

Looming like a giant flying saucer in our outer solar system, Saturn puts on a show as the planet and its magnificent ring system nod majestically over the course of its 29-year journey around the Sun. These Hubble Space Telescope images, captured from 1996 to 2000, show Saturn's rings open up from just past edge-on to nearly fully open as it moves from autumn towards winter in its Northern Hemisphere.

Astronomers are studying this set of images to investigate the detailed variations in the color and brightness of the rings. They hope to learn more about the rings' composition, how they were formed, and how long they might last. Saturn's rings are incredibly thin, with a thickness of only about 30 feet (10 meters). The rings are made of dusty water ice, in the form of boulder-sized and smaller chunks that gently collide with each other as they orbit around Saturn. Saturn's gravitational field constantly disrupts these ice chunks, keeping them spread out and preventing them from combining to form a moon. The rings, as shown here, have a slight pale reddish color due to the presence of organic material mixed with the water ice.

The first image in this sequence, on the lower left, was taken soon after the autumnal equinox in Saturn's Northern Hemisphere (which is the same as the spring equinox in its Southern Hemisphere). By the final image in the sequence, on the upper right, the tilt is nearing its extreme, or winter solstice in the Northern Hemisphere (summer solstice in the Southern Hemisphere).

Careers in Astronomy

American Astronomical Society, 2000 Florida Ave., N.W., Suite 400, Washington, DC, 20009; aas@aas.org; www.aas.org. Information on careers in astronomy is available on this Society's website.

History

James A. Connor, *Kepler's Witch* (San Francisco: HarperSanFrancisco, 2004).

Stillman Drake, *Galileo: A Very Short Introduction* (New York: Oxford University Press, 2001).

Kitty Ferguson, *The Nobleman and His Housedog: Tycho Brahe and Johannes Kepler: The Strange Partnership that Revolutionized Science* (Walker & Co, 2002). http://tychoandkepler.com.

Owen Gingerich and James MacLachlan, *Nicolaus Copernicus: Making the Earth a Planet* (New York: Oxford University Press, 2005).

Owen Gingerich, *The Book Nobody Read: Chasing the Revolutions of Nicolaus Copernicus* (Walker and Co., 2004, paperback 2005). About the hunt for copies of Copernicus's book.

Owen Gingerich, *The Eye of Heaven: Ptolemy, Copernicus and Kepler* (New York: American Institute of Physics, 1993).

Michael Hoskin, ed., *The Cambridge Concise History of Astronomy* (Cambridge University Press, 1999).

Michael Hoskin, ed., *Cambridge Illustrated History: Astronomy* (Cambridge University Press, 1997).

Rocky Kolb, *Blind Watchers of the Sky* (Addison-Wesley, 1996). The evolution of world views, beginning with Tycho Brahe.

James MacLachlan, *Galileo Galilei: First Physicist* (New York: Oxford University Press, 1999).

Stephen P. Maran and Laurence A. Marschall, *Galileo's New Universe: The Revolution in Our Understanding of the Cosmos* (BenBella Books, 2009).

James Reston, *Galileo: A Life* (New York: HarperCollins, 1994).

Dava Sobel, *Galileo's Daughter: A Historical Memoir of Faith, Science, and Love* (Walker & Co., 1999, 2000).

James Voelkel, *Johannes Kepler and the New Astronomy* (New York: Oxford University Press, 1999). A short biography in the Oxford Portraits series.

Art and Astronomy

Roberta J. M. Olson and Jay M. Pasachoff, *Fire in the Sky: Comets and Meteors, the Decisive Centuries, in British Art and Science* (Cambridge University Press, 1998, 1999). From the time of Newton and Halley to the present.

Solar System

Walter Alvarez, *T.-Rex and the Crater of Doom* (Princeton: Princeton Univ. Press, 1997). A personal account of the quest to understand the extinction of the dinosaurs.

Donald A. Beattie, *Taking Science to the Moon* (Johns Hopkins University Press).

J. Kelly Beatty, Caroline Collins Petersen, and Andrew Chaikin, *The New Solar System*, 4th ed. (Cambridge, MA: Sky Publishing Corp. and Cambridge University Press, 1999). Each chapter written by a different expert.

Reta Beebe, *Jupiter: The Giant Planet* (Washington, D.C.: Smithsonian Institution Press, 2nd ed., 1996).

Jim Bell and Jacqueline Mitton, eds., *Asteroid Rendezvous: NEAR Shoemaker's Adventures at Eros* (Cambridge University Press, 2002).

Joseph M. Boyce, *The Smithsonian Book of Mars* (Smithsonian Institution Press, 2002).

Michael Brown, *How I Killed Pluto and Why It Had It Coming* (Random House, 2010).

Eugene Cernan and Don David, *The Last Man on the Moon* (New York: St. Martin's Press, 1999). A first-person account.

Andrew Chaikin, *A Man on the Moon: The Voyages of the Apollo Astronauts* (New York: Viking, 1994). The story of the missions and the people on them.

Clark Chapman and David Morrison, *Cosmic Catastrophes* (Plenum, 1989). An interesting account of the many ways in which life on Earth is threatened.

Neil Comins, *What if the Moon Didn't Exist?: Voyages to Earths that Might Have Been* (HarperCollins, 1993).

Neil Comins, *What if the Earth Had Two Moons?* (St. Martin's Press, 2010).

Ken Croswell, *Planet Quest: The Epic Discovery of Alien Solar Systems* (New York: Free Press, 1997). Excellent early account of the search for extrasolar planets.

Imke de Pater and Jack J. Lissauer, *Planetary Sciences* (Cambridge University Press, 2001).

James Elliot and Richard Kerr, *Rings* (Cambridge, MA: MIT Press, 1984). Includes first-person and other stories of the discoveries.

Fred Espenak, *Fifty Year Canon of Lunar Eclipses: 1968–2035* (NASA Ref. Pub. 1216).

Tim Ferris, *Seeing in the Dark: How Backyard Stargazers Are Probing Deep Space and Guarding Earth from Interplanetary Peril* (Simon & Schuster, 2002).

Daniel Fischer, *Mission Jupiter: The Spectacular Journey of the Galileo Spacecraft* (New York: Copernicus, 2001).

M. Garlick, *The Story of the Solar System* (Cambridge University Press, 2002).

Donald Goldsmith and Tobias Owen, *The Search for Life in the Universe* (Addison-Wesley, 1992). Describes the conditions thought to be necessary for life as we know it.

Donald Goldsmith, *The Hunt for Life on Mars* (Dutton, 1997).

Donald W. Goldsmith, *Worlds Unnumbered: The Quest to Discover Other Solar Systems* (Sausalito, CA: University Science Books, 1997).

Richard Greenberg, *Unmasking Europa: The Search for Life on Jupiter's Ocean Moon* (New York: Springer Praxis Books, 2008).

David M. Harland, *Exploring the Moon: The Apollo Expeditions* (New York: Springer-Verlag, 1999).

Paul Hodge, *Higher than Everest: An Adventurer's Guide to the Solar System* (Cambridge University Press, 2001).

Jeffrey S. Kargel, *Mars—A Warmer, Wetter Planet* (New York: Springer Praxis Books, 2004).

Jeffrey Kluger, *Moon Hunters: NASA's Remarkable Expeditions to the Ends of the Solar System* (Simon and Schuster, 2001).

Gary W. Kronk, *Cometography: A Catalogue of Comets, vol. 1, Ancient–1799; vol. 2, 1800–1899; vol. 3, 1900–1943; vol. 4, 1933–1959* (Cambridge University Press, 1999, 2003, 2007, 2009, respectively).

Kenneth Lang, *The Cambridge Guide to the Solar System* (Cambridge University Press, 2003).

Michael Lemonick, *Other Worlds: The Search for Life in the Universe* (New York: Simon & Schuster, 1998).

Eli Maor, *June 8, 2004: Venus in Transit* (Princeton University Press, 2000).

Michael Maunder and Patrick Moore, *Transit: When Planets Cross the Sun* (New York: Springer-Verlag, 2000).

Ron Miller & William K. Hartmann, *The Grand Tour: A Traveler's Guide to the Solar System*, 2005 revised edition.

David McNab and James Younger, *The Planets* (Yale University Press, 1999). To accompany a TV series on planetary exploration.

David Morrison, *Exploring Planetary Worlds* (Scientific American Library, 1993).

David Morrison and Tobias Owen, *The Planetary System*, 2nd ed. (Reading, MA: Addison-Wesley, 1996). All about the solar system.

J. H. Rogers, *The Giant Planet Jupiter* (Cambridge University Press, 1995).

Carl Sagan, *Pale Blue Dot* (New York: Random House, 1994). One of astronomy's most eloquent spokeperson's last works.

William Sheehan and John Westfall, *The Transits of Venus* (Prometheus Books, 2004).

Steven Squyres, *Roving Mars: Spirit, Opportunity, and the Exploration of the Red Planet* (2005). By the principal investigator of the Mars Exploration Rovers Spirit and Opportunity.

S. Alan Stern, *Our worlds: The magnetism and thrill of planetary exploration: as described by leading planetary scientists* (Cambridge University Press, 1999).

S. Alan Stern and Jacqueline Mitton, *Pluto & Charon: Ice Worlds on the Ragged Edge of the Solar System* (New York: Wiley, 1997).

Reginald Turnhill, *The Moon Landings* (Cambridge University Press, 2002).

Peter Douglas Ward and Donald Brownlee. *Rare Earth: Why Complex Life is Uncommon in the Universe* (New York: Copernicus Books, 2000). Argues that intelligent, technologically advanced life is very rare in the cosmos.

Don E. Wilhelms, *To a Rocky Moon: A Geologist's History of Lunar Exploration* (Tucson: Univ. of Arizona Press, 1993).

Ben Zuckerman and Michael Hart, eds., *Extraterrestrials: Where Are They?* 2nd ed. (Cambridge University Press, 1995).

Pluto and Dwarf Planets

Neil DeGrasse Tyson, *The Pluto Files*, 2008 (NOVA in March 2010 on PBS)

Alan Boyle, *The Case for Pluto: How a Little Planet Made a Big Difference*, 2010.

Michael Brown, to be published by Random House in fall 2010.

Laurence A. Marschall & Stephen P. Maran, *Pluto Confidential: An Insider Account of the Ongoing Battles Over the Status of Pluto*, 2009.

Marc McCutcheon, *The Kid Who Named Pluto*, 2004, pages 20–27.

Govert Schilling, *The Hunt for Planet X: New Worlds and the Fate of Pluto* (New York: Springer, 2009).

The Sun

Arvind Bhatnagar and William C. Livingston, *Fundamentals of Solar Astronomy* (World Scientific, 2005). Comprehensive and phenomenological but relatively non-mathematical.

Michael J. Carlowicz and Ramon E. Lopez, *Storms from the Sun: The Emerging Science of Space Weather* (Joseph Henry Press, 2000).

Fred Espenak, *Fifty Year Canon of Solar Eclipses* (NASA Ref. Pub. 1178, Rev. 1987). Maps and tables.

Leon Golub and Jay M. Pasachoff, *Nearest Star: The Surprising Science of Our Sun* (Harvard University Press, 2001). A non-technical trade book.

Leon Golub and Jay M. Pasachoff, *The Solar Corona*, 2nd ed. (Cambridge Univ. Press, 2009). An advanced text.

Kenneth R. Lang, *Sun, Earth, and Sky*, 2nd ed. (Springer-Verlag, 2006).

Kenneth J. H. Phillips, *Guide to the Sun* (New York: Cambridge Univ. Press, 1992).

Peter O. Taylor and Nancy L. Hendrickson, *Beginner's Guide to the Sun* (Waukesha, WI, Kalmbach Books, 1995).

Jack B. Zirker, *Journey From the Center of the Sun* (Princeton University Press, 2001, paperback, 2004). A non-technical trade book.

Jack B. Zirker, *Sunquakes: Probing the Interior of the Sun* (Princeton University Press, 2003). A non-technical trade book.

Jack B. Zirker, *The Magnetic Universe: The Elusive Traces of an Invisible Force* (Johns Hopkins University Press, 2009). A non-technical trade book.

INDEX

The Smart Guide Series

Making Smart People Smarter

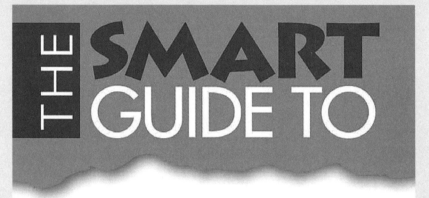

GREEN LIVING

The most complete guide to green living ever published

How green living benefits your health as well as the Earth's

How green living can save you lots of money

Why the green economy and job market is an attractive, new, lucrative frontier

Julie Kerr **Gines**

Titles

Smart Guide To Bachelorette Parties
Smart Guide To Biology
Smart Guide To Bridge
Smart Guide To Chemistry
Smart Guide To Classical Music
Smart Guide To eBay
Smart Guide To Fighting Infections
Smart Guide To Freshwater Fishing
Smart Guide To Getting Published
Smart Guide To Green Living
Smart Guide To Healthy Grilling
Smart Guide To High School Math
Smart Guide To Hiking and Backpacking
Smarat Guide To The History of Science
Smart Guide To Horses and Riding
Smart Guide To Life After Divorce
Smart Guide To Managing Stress
Smart Guide To Nutrition
Smart Guide To Patents
Smart Guide To Practical Math
Smart Guide To Single Malt Scotch Whiskey
Smart Guide To Starting Your Own Business
Smart Guide To The Perfect Job Interview
Smart Guide To The Solar System
Smart Guide To Understanding Your Cat
Smart Guide To US Visas
Smart Guide To Wedding Weekend Events
Smart Guide To Wine

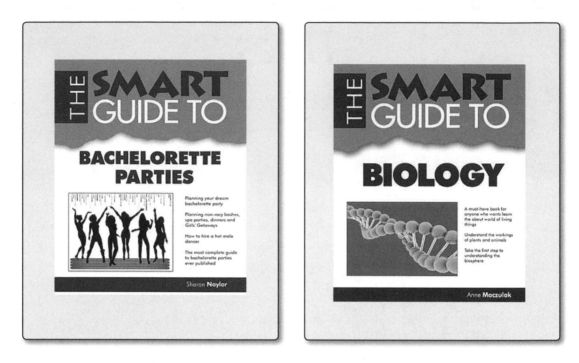

THE SMART GUIDE TO BACHELORETTE PARTIES

Planning your dream bachelorette party

Planning non-racy bashes, spa parties, dinners and Girls' Getaways

How to hire a hot male dancer

The most complete guide to bachelorette parties ever published

Sharon **Naylor**

THE SMART GUIDE TO BIOLOGY

A must-have book for anyone who wants learn the about world of living things

Understand the workings of plants and animals

Take the first step to understanding the biosphere

Anne **Maczulak**

THE SMART GUIDE TO CLASSICAL MUSIC

A must-have book for anyone who wants to fully understand classical music

Quick and easy guidance for getting the most out of a recording, broadcast or concert

Down-to-earth advice to help you make CD selections

Robert **Sherman** and Philip **Seldon**

THE SMART GUIDE TO MASTERING EBAY

A must-have book for anyone who wants to make money on eBay

Buying and selling on eBay like a pro

Building your own eBay store

Get a complete walkthrough of eBay

Nieves **Marcus**

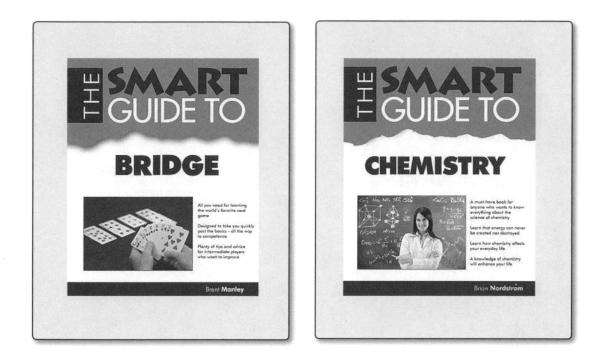

THE SMART GUIDE TO

HIKING AND BACKPACKING

A must-have book for anyone who enjoys getting into the outdoors

Useful information for both beginning and experienced hikers and backpackers

Learn how to stay safe and healthy while hiking and backpacking

Brian **Nordstrom**

THE SMART GUIDE TO

HISTORY OF SCIENCE

One of the most complete guides ever to the worldwide history of science

Covers the scientific history of both the Eastern and Western hemispheres

Is the concise, fun-to-read outcome of mankind's insatiable quest for knowledge

Julie Kerr **Gines**

THE SMART GUIDE TO

MANAGING STRESS

Tame your worries and fears

Handle unreasonable people who stress you out

Reduce tension with loved ones and co-workers

Tips to beef up your resistance to stress

Bryan **Robinson**

THE SMART GUIDE TO

NUTRITION

Understanding the balanced diet

Learn the truth about sugar

Learn how to spot harmful fad diets

Managing salt intake and other nutrition pitfalls

How to select nutrient-rich foods

Anne **Maczulak**

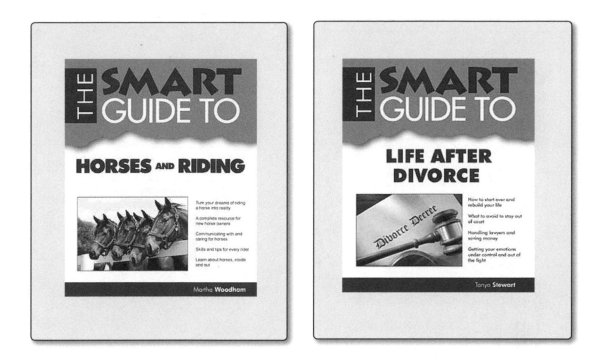

THE SMART GUIDE TO

HORSES AND RIDING

Turn your dreams of riding a horse into reality

A complete resource for new horse owners

Communicating with and caring for horses

Skills and tips for every rider

Learn about horses, inside and out

Martha **Woodham**

THE SMART GUIDE TO

LIFE AFTER DIVORCE

How to start over and rebuild your life

What to avoid to stay out of court

Handling lawyers and saving money

Getting your emotions under control and out of the fight

Tanya **Stewart**

THE SMART GUIDE TO

PATENTS

How to protect your idea from the first moment

Getting the best possible patent on your idea

Turning a patent into a source of income

Taking the steps now so you can win big later if someone infringes

Aaron **Filler**

THE SMART GUIDE TO

PRACTICAL MATH

Answers to the problems you need to solve - in an easy-to-use format

Practical math problems for your home, your finances, your health, your business

Practical solutions that give you the fastest, cheapest, or easiest way to do a job

Jim **Stein**

THE SMART GUIDE TO
SINGLE MALT SCOTCH WHISKY

A must-have book for anyone who wants to know everything about single malt Scotch whisky

Information about all the distilleries and the whisky they make

Learn how to taste and appreciate single malt Scotch Whisky

Elizabeth Riley **Bell**

THE SMART GUIDE TO
STARTING YOUR OWN BUSINESS

A must-have book for anyone who wants to start their own business

Learn how to find startup and backup capital

Cheap marketing ideas and using publicity

First year problems and how to solve them

Barry **Thomsen**

THE SMART GUIDE TO
UNDERSTANDING YOUR CAT

If only cats could talk, they'd tell us almost everything in this book

How to connect with your cat

Everything you need to know to make your cat feel at home in your home

Take a closer look at your lovable feline friend

Carolyn **Janik**

THE SMART GUIDE TO
UNITED STATES VISAS

What a visa is and how you get one

The difference between a visa and a greencard

How to navigate the byzantine process of applying for a visa

The various visa categories and what they mean

Scott **Syfert** & Melisa **Boris**